Homeschool Testing Book
for
Algebra 2
Third Edition

by
John H. Saxon Jr.

CONTENTS

SAXON™
A Harcourt Achieve Imprint

www.SaxonHomeschool.com
1.800.416.8171

Test Solutions

About the Solutions

These solutions are designed to be representative of students' work. Please keep in mind that many problems will have more than one correct solution. We have attempted to stay as close as possible to the methods and procedures outlined in the textbook. The final answers are set in boldface for ease of grading. Below each problem number are one or more small numbers in parentheses called Lesson Reference Numbers. These numbers refer to the lessons in which the concepts for a particular problem were taught.

TEST 1

1.
(A)
$$m + 81 = 180$$
$$m = \mathbf{99}$$

Since vertical angles are equal:
$$n = \mathbf{99}$$
$$3p = 81$$
$$p = \mathbf{27}$$

2.
(A)
$$3x + 120 = 180$$
$$3x = 60$$
$$x = \mathbf{20}$$

Since vertical angles are equal:
$$6y = 120$$
$$y = \mathbf{20}$$
$$4z = 60$$
$$z = \mathbf{15}$$

3.
(A)
$$x + 37° = 90°$$
$$x = \mathbf{53°}$$

4.
(B)
$$P = \frac{1}{2}(2)(\pi)(3) + \frac{1}{2}(2)(\pi)(4) + 48$$
$$= 3\pi + 4\pi + 48$$
$$= 7\pi + 48 \approx \mathbf{69.98\ in.}$$

5.
(B)
$$A = \frac{60}{360} \cdot \pi(8)^2$$
$$= \frac{60}{360} \cdot 64\pi \approx \mathbf{33.49\ in.^2}$$

6.
(B)
$$A_{\text{Base}} = \frac{1}{2}\pi(3)^2 + (18)(8)$$
$$= \frac{9}{2}\pi + 144 \approx 158.13\ m^2$$
$$V = \frac{1}{3}A_{\text{Base}} \times H$$
$$\approx \frac{1}{3}(158.13)(4) \approx \mathbf{210.84\ m^3}$$

7.
(B)
$$A = \frac{1}{2}(12)(8) - \frac{1}{2}(5)(2) - \frac{1}{2}(10)(3) - (2)(3)$$
$$= 48 - 5 - 15 - 6 = \mathbf{22\ ft^2}$$

8.
(B)
$$A = 12(5) - \pi(2)^2 = 60 - 4\pi \approx \mathbf{47.44\ m^2}$$

9.
(B)
$$A_{\text{Base}} = \pi(5)^2 = 25\pi\ in.^2$$
$$V = A_{\text{Base}} \times H$$
$$750\pi = (25\pi)H$$
$$H = \mathbf{30\ in.}$$

10.
(1)
Since angles opposite equal sides are equal angles,
$$a = \mathbf{42}.$$
$$b + 42 + 42 = 180$$
$$b = \mathbf{96}$$

11.
(1)
$$4 \times \overrightarrow{SF} = 7$$
$$\overrightarrow{SF} = \frac{7}{4}$$
$$3 \times \overrightarrow{SF} = y$$
$$3\left(\frac{7}{4}\right) = y$$
$$y = \frac{\mathbf{21}}{\mathbf{4}}$$

12.
(B)
$$V = \frac{2}{3}A_{\text{Base}} \times \text{height}$$
$$= \frac{2}{3}\left[\pi(8)^2\right](16)$$
$$= \frac{2}{3}(1024\pi) \approx \mathbf{2143.57\ cm^3}$$
$$S.A. = 4\pi r^2 = 4\pi(8)^2 \approx \mathbf{803.84\ cm^2}$$

13.
(2)
$$\frac{(xy^2)^0\, x^2 y}{x(y^{-3})^3} = \frac{x^2 y}{xy^{-9}} = \mathbf{xy^{10}}$$

14.
(2)
$$\frac{(x^3 y^{-1})^{-2}\, z^{-2}}{(y^3 z y^{-2})^5} = \frac{x^{-6} y^2 z^{-2}}{y^{15} z^5 y^{-10}} = \mathbf{x^{-6} y^{-3} z^{-7}}$$

15.
(2)
$$\frac{x^3 y^2 z^{-2}}{(xw^0)^{-2}\, z^{-1} x^2 w^3} = \frac{x^3 y^2 z^{-2}}{x^{-2} z^{-1} x^2 w^3}$$
$$= \mathbf{x^3 y^2 z^{-1} w^{-3}}$$

16.
(2)
$$-3^{-5} = -\frac{1}{3^5} = -\frac{\mathbf{1}}{\mathbf{243}}$$

17.
(2)
$$\frac{1}{-3^{-3}} = -3^3 = \mathbf{-27}$$

18.
(A)
$$-4^3 - [-5^0 - (3 - 5) - 4]$$
$$= -64 - [-1 + 2 - 4] = -64 - [-3] = \mathbf{-61}$$

19.
(A)
$$-|-3 - 5| - (-3)^2 - 3^2 = -|-8| - 9 - 9$$
$$= -8 - 9 - 9 = \mathbf{-26}$$

20.
(A)
$$-3[-6^0 - 2(6 - 8) - 2^3]$$
$$= -3[-1 - 2(-2) - 8]$$
$$= -3[-1 + 4 - 8] = -3[-5] = \mathbf{15}$$

TEST 2

1.
(1)

$$8 \times \overrightarrow{SF} = \frac{29}{3}$$

$$\overrightarrow{SF} = \frac{29}{24}$$

$$15 \times \overrightarrow{SF} = y$$

$$15\left(\frac{29}{24}\right) = y$$

$$y = \frac{145}{8}$$

Since lines are parallel:

$$3A + 120 = 180 \qquad 5B = 120$$
$$3A = 60 \qquad B = 24$$
$$A = 20$$

2.
(B)

$$A_{\text{Base}} = \frac{1}{2}(3)(4) + \frac{40}{360}(\pi)(4)^2$$

$$= 6 + \frac{40}{360}(16\pi) \approx 11.58 \text{ cm}^2$$

$$V = \frac{1}{3}A_{\text{Base}} \times H$$

$$\approx \frac{1}{3}(11.58)(12) \approx 46.32 \text{ cm}^3$$

3.
(1)

$$x + y + 70 = 180$$
$$x + y = 110$$

Since angles opposite equal sides are equal angles,
$x = y = 55$.

$$x + P = 180$$
$$55 + P = 180$$
$$P = 125$$

$$Q + R + P = 180$$
$$Q + R + 125 = 180$$
$$Q + R = 55$$

Since angles opposite equal sides are equal angles,

$$Q = R = \frac{55}{2} = 27.5.$$

4.
(B)

$$\text{S.A.} = 4\pi r^2 = 38\pi$$

$$r = \sqrt{\frac{38\pi}{4\pi}}$$

$$r = \frac{\sqrt{38}}{2} \text{ in.}$$

5.
(3)

$$a - |b|a^2 - ba = -3 - |-4|(-3)^2 - (-4)(-3)$$
$$= -3 - 4(9) - 12 = -51$$

6.
(3)

$$m^2 n(mn + n^2)$$

$$= \left(-\frac{1}{2}\right)^2\left(\frac{1}{4}\right)\left[\left(-\frac{1}{2}\right)\left(\frac{1}{4}\right) + \left(\frac{1}{4}\right)^2\right]$$

$$= \frac{1}{4}\left(\frac{1}{4}\right)\left[-\frac{1}{8} + \frac{1}{16}\right] = \frac{1}{16}\left[-\frac{1}{16}\right] = -\frac{1}{256}$$

7.
(3)

$$2ab^{-1} + \frac{4a^2 y^{-1}}{a} - \frac{7x^{-1}y}{y^{-2}}$$

$$= 2ab^{-1} + 4ay^{-1} - 7x^{-1}y^3$$

8.
(3)

$$-2mn + \frac{3mn^{-2}m^0}{n^{-3}} - \frac{5m^2 m^{-1}m}{(m^{-1})^{-1}}$$

$$= -2mn + 3mn - 5m = mn - 5m$$

9.
(2)

$$\frac{xb^0 y(x^{-2}b^{-2})^2}{xb(by^0)xby} = \frac{xyx^{-4}b^{-4}}{x^2 b^3 y} = x^{-5}b^{-7}$$

10.
(2)

$$\frac{4(a^{-2})^{-1}b^2 a^4 b^{-2}}{a^2 aa^0 a^{-5}(a^3)^2} = \frac{4a^2 a^4}{a^3 a^{-5} a^6} = 4a^2$$

11.
(A)

$$-5^0[-3^3 - 3(-3 - 2)](-4^0)$$

$$= -1[-27 - 3(-5)](-1)$$

$$= -1[-12](-1) = -12$$

12.
(2)

$$-3^{-3} - \frac{4}{-4^{-2}} - 2^0 = -\frac{1}{3^3} + 4^3 - 1$$

$$= -\frac{1}{27} + 64 - 1 = 62\frac{26}{27}$$

13.
(A)

$$-4^2 + (-2)^4 - 3^3 - |-4 - 4|$$

$$= -16 + 16 - 27 - 8 = -35$$

14.
(6)

$$0.003x + 0.6 = 2.88$$
$$3x + 600 = 2880$$
$$3x = 2280$$
$$x = 760$$

15. $4\frac{2}{3}x - 2\frac{1}{5} = 3\frac{1}{6}$
(4)

$$\frac{14}{3}x = \frac{19}{6} + \frac{11}{5}$$

$$\frac{14}{3}x = \frac{95}{30} + \frac{66}{30}$$

$$\frac{14}{3}x = \frac{161}{30}$$

$$x = \frac{161}{30} \cdot \frac{3}{14} = \frac{23}{20} = 1\frac{3}{20}$$

16. $-2 - 2^3 - 2(x - 2) = 2[(x - 2)2 - 2]$
(4)

$$-2 - 8 - 2x + 4 = 2[2x - 6]$$

$$-2x - 6 = 4x - 12$$

$$6x = 6$$

$$x = 1$$

17. $\frac{ab}{m}\left(\frac{-2m^{-2}}{ba} + \frac{3m}{a^{-1}b}\right) = \frac{-2abm^{-2}}{mba} + \frac{3abm}{ma^{-1}b}$
(4)

$$= -2m^{-3} + 3a^2$$

18. $x = 4(90 - x)$
(5)

$$x = 360 - 4x$$

$$5x = 360$$

$$x = 72°$$

19. $N \qquad N + 2 \qquad N + 4$
(6)

$$3(N + N + 4) = 4(N + 2) + 18$$

$$6N + 12 = 4N + 26$$

$$2N = 14$$

$$N = 7$$

The desired integers are **7, 9,** and **11.**

20. $1 - 0.382 = 0.618$
(6)

If 0.382 are totally loyal, then 0.618 are not totally loyal.

$$WD \times of = is$$

$$0.618(8000) = NL$$

$$NL = \textbf{4944 employees}$$

TEST 3

1. $\frac{P}{100} \times of = is \quad \longrightarrow \quad \frac{30}{100} \times WN = 480$
(7)

$$WN = 480 \cdot \frac{100}{30} = \textbf{1600}$$

Since one part of 1600 is 480 for 30%, the other part must be 1120 for 70%.

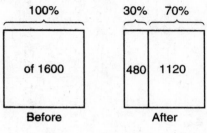

2.
(9)

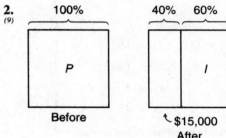

$$\frac{P}{100} \times of = is \quad \longrightarrow \quad \frac{40}{100} \times P = \$15,000$$

$$P = \$15,000 \cdot \frac{100}{40} = \textbf{\$37,500}$$

3. $N \qquad N + 2 \qquad N + 4$
(6)

$$3N = 2(N + 2 + N + 4) - 26$$

$$3N = 4N - 14$$

$$N = 14$$

The desired integers are **14, 16,** and **18.**

4. $F \times of = is$
(5)

$$2\frac{3}{4} \times R = 3300$$

$$R = 3300 \cdot \frac{4}{11} = \textbf{1200 walk randomly}$$

5. Since angles opposite equal sides are equal angles,
(1) $b = \textbf{75.}$

$$a + b + 75 = 180$$

$$a = 180 - 75 - 75 = \textbf{30}$$

$$b + \angle XYZ = 180$$

$$\angle XYZ = 180 - 75 = 105$$

$$c + 35 + 105 = 180$$

$$c = 180 - 105 - 35 = \textbf{40}$$

6. Since lines are parallel:
(7)

$$(4x + 7) + (3x - 16) = 180$$

$$7x - 9 = 180$$

$$7x = 189$$

$$x = \textbf{27}$$

7. $4 \times \overleftrightarrow{SF} = \dfrac{7}{2}$
(7)

$\overleftrightarrow{SF} = \dfrac{7}{2} \cdot \dfrac{1}{4} = \dfrac{7}{8}$

$6 \times \overleftrightarrow{SF} = a$

$6\left(\dfrac{7}{8}\right) = a$

$\dfrac{21}{4} = a$

Since lines are parallel:

$(5x - 19) + (3x - 1) = 180$

$8x - 20 = 180$

$8x = 200$

$x = 25$

Since vertical angles are equal angles:

$R = 5x - 19 = 5(25) - 19 = \mathbf{106}$

$T = 3x - 1 = 3(25) - 1 = \mathbf{74}$

8. $c^2 = a^2 + b^2$
(10)

$16^2 = m^2 + 10^2$

$256 = m^2 + 100$

$156 = m^2$

$m = \sqrt{156} = \mathbf{2\sqrt{39}}$

9. $2\dfrac{1}{4}n - \dfrac{5}{12} = \dfrac{3}{4}$
(4)

$\dfrac{9}{4}n = \dfrac{3}{4} + \dfrac{5}{12}$

$\dfrac{9}{4}n = \dfrac{9}{12} + \dfrac{5}{12}$

$\dfrac{9}{4}n = \dfrac{14}{12}$

$n = \dfrac{14}{12} \cdot \dfrac{4}{9} = \mathbf{\dfrac{14}{27}}$

10. $-3(2x - 1) - 3^2 + 7^0 = -3(3x + 2x^0)$
(4)

$-6x + 3 - 9 + 1 = -9x - 6$

$-6x - 5 = -9x - 6$

$3x = -1$

$x = -\dfrac{1}{3}$

11. $\dfrac{p^{-2}c^3}{c^{-2}}\left(2p^2c - \dfrac{3pcp^2}{pc^2p}\right) = \dfrac{2c^4}{c^{-2}} - \dfrac{3p^{-2}c^4p^3}{p^2}$
(4)

$= \mathbf{2c^6 - 3p^{-1}c^4}$

12. $\dfrac{ac^2}{x}\left(\dfrac{3a^{-1}x}{c^2} - \dfrac{7cx}{a}\right) = \dfrac{3c^2x}{xc^2} - \dfrac{7ac^3x}{xa}$
(4)

$= \mathbf{3 - 7c^3}$

13. $A = \pi r^2 = 25\pi$
(2)

$r = \sqrt{\dfrac{25\pi}{\pi}} = 5$

$C = 2\pi r = 2\pi(5) = \mathbf{10\pi \ m}$

14. $2x - y = 2$
(8)

$-y = -2x + 2$

$y = 2x - 2$

The y-intercept is -2; the slope is 2.

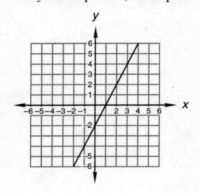

15. $xy - x(y - x) = -\dfrac{1}{2}\left(-\dfrac{1}{3}\right) - \left(-\dfrac{1}{2}\right)\left(-\dfrac{1}{3} + \dfrac{1}{2}\right)$
(3)

$= \dfrac{1}{6} + \dfrac{1}{2}\left(\dfrac{1}{6}\right) = \dfrac{1}{6} + \dfrac{1}{12} = \mathbf{\dfrac{1}{4}}$

16. $a^2c(a + c) - c^0 = \left(-\dfrac{1}{3}\right)^2\left(-\dfrac{1}{2}\right)\left(-\dfrac{1}{3} - \dfrac{1}{2}\right) - 1$
(3)

$\dfrac{1}{9}\left(-\dfrac{1}{2}\right)\left(-\dfrac{5}{6}\right) - 1 = \dfrac{5}{108} - 1 = \mathbf{-\dfrac{103}{108}}$

17. $\dfrac{2^{-3}aaa^0(x^{-3}c)^{-2}}{(a^3c^{-1})^2acc^{-2}} = \dfrac{a^2x^6c^{-2}}{8a^6c^{-2}ac^{-1}} = \mathbf{\dfrac{x^6c}{8a^5}}$
(2)

18. $-4^0(-3^0 - 5^0 - |-3|) - (-3)(-6)$
(A)

$= -1(-1 - 1 - 3) - 18 = 5 - 18 = \mathbf{-13}$

19. $-\dfrac{1}{3^{-2}} - 2^3 - 2 = -3^2 - 8 - 2$
(2)

$= -9 - 8 - 2 = \mathbf{-19}$

20. $\dfrac{5c^2y}{ac} + \dfrac{2c}{a} - \dfrac{4y^3c}{cya} - \dfrac{2c^3a}{a^2c^2}$
(3)

$= \dfrac{5cy}{a} + \dfrac{2c}{a} - \dfrac{4y^2}{a} - \dfrac{2c}{a} = \mathbf{\dfrac{5cy}{a} - \dfrac{4y^2}{a}}$

TEST 4

1.
(9)
$$\frac{P}{100} \times of = is$$

$$\frac{340}{100} \times B = 68{,}000$$

$$B = 68{,}000 \cdot \frac{100}{340} = \textbf{20,000 flowers}$$

2.
(5)
$$-3(2N + 6) = 3(-N) + 54$$
$$-6N - 18 = -3N + 54$$
$$-3N = 72$$
$$N = \textbf{–24}$$

3.
(6)
$$N \qquad N + 2 \qquad N + 4$$
$$6N = 4(N + 2 + N + 4) - 16$$
$$6N = 8N + 8$$
$$2N = -8$$
$$N = -4$$

The desired integers are **–4, –2,** and **0.**

4.
(11)
$$c = \frac{1}{2}(130) = \textbf{65}$$

$$(3d - 7) + (4d + 13) = 360 - 130$$
$$7d + 6 = 230$$
$$7d = 224$$
$$d = \textbf{32}$$

5.
(1)
$$(180 - a) + 70 + (180 - 150) = 180$$
$$-a + 180 + 70 + 30 = 180$$
$$a = \textbf{100}$$

6.
(13)
$$15^2 = H^2 + 10^2$$
$$225 = H^2 + 100$$
$$125 = H^2$$
$$5\sqrt{5} = H$$

$$A = \frac{B \times H}{2} = \frac{(20 \times 5\sqrt{5})}{2} = \textbf{50}\sqrt{\textbf{5}} \textbf{ cm}^2$$

7.
(12)
(a) Every point is 2 units below the x-axis.
$$y = \textbf{–2}$$

(b) The y-intercept is +1. The slope is negative and the rise over the run for any triangle drawn is $-\frac{4}{3}$.

$$y = -\frac{4}{3}x + 1$$

8.
(11)
$$3 + \frac{r}{3t^2} = \frac{9t^2}{3t^2} + \frac{r}{3t^2} = \frac{9t^2 + r}{3t^2}$$

9.
(11)
$$\frac{a}{py} + y + \frac{a^2}{p^2} = \frac{ap}{p^2y} + \frac{p^2y^2}{p^2y} + \frac{a^2y}{p^2y}$$

$$= \frac{ap + p^2y^2 + a^2y}{p^2y}$$

10.
(10)
$$D^2 = 3^2 + 9^2$$
$$D^2 = 9 + 81$$
$$D^2 = 90$$
$$D = \sqrt{90}$$
$$D = \textbf{3}\sqrt{\textbf{10}}$$

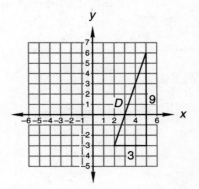

11.
(8)
(a) $x = 3$

(b) $3x - 4y = 12$
$$-4y = -3x + 12$$
$$y = \frac{3}{4}x - 3$$

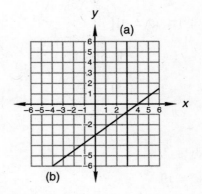

12.
(6)
$$0.3x - 0.3 - 0.03 = 0.33$$
$$30x - 30 - 3 = 33$$
$$30x = 66$$
$$x = \textbf{2.2}$$

13.
(4)
$$-2^0(m^0 - 2) - 3(m - 3) = -2(m + 3^0)$$
$$-1 + 2 - 3m + 9 = -2m - 2$$
$$-3m + 10 = -2m - 2$$
$$m = \textbf{12}$$

14.
(4)

$$\frac{3^{-2} x^{-2}}{y}\left(3x^2 y - \frac{9x^2}{y^2}\right)$$

$$= \frac{3^{-2} x^{-2}\, 3x^2 y}{y} - \frac{3^{-2} x^{-2}\, 9x^2}{yy^2}$$

$$= 3^{-1} - \frac{1}{y^3} = \mathbf{3^{-1} - y^{-3}}$$

15.
(2)

$$\frac{4^{-2}\, mmm^0\, (m^{-3}p)^{-3}}{(m^3 p^{-1})^3\, mpp^{-4}} = \frac{m^2 m^9 p^{-3}}{16 m^9 p^{-3} mp^{-3}}$$

$$= \frac{m^{11} p^{-3}}{16 m^{10} p^{-6}} = \mathbf{\frac{mp^3}{16}}$$

16.
(3)

$$-\frac{3pq}{r} + \frac{5p^4 p^{-3} r^{-1}}{q} - \frac{2r^{-1} p^{-2}}{p^{-3} q^{-1}}$$

$$= -\frac{3pq}{r} + \frac{5p}{qr} - \frac{2pq}{r} = \mathbf{-\frac{5pq}{r} + \frac{5p}{qr}}$$

17.
(3)

$$a^2 - ya - y(a - y)$$

$$= \left(\frac{1}{3}\right)^2 - (-3)\left(\frac{1}{3}\right) - (-3)\left(\frac{1}{3} + 3\right)$$

$$= \frac{1}{9} + 1 + 1 + 9 = \mathbf{11\frac{1}{9}}$$

18.
(A)

$$5^0 - \frac{75}{5^2} + (-4)^0 - 4^0$$

$$- 3\big[(-5 - 4^0) - (2 - 3)\big]$$

$$= 1 - \frac{75}{25} + 1 - 1 - 3[-6 + 1]$$

$$= 1 - 3 + 1 - 1 - 3[-5] = \mathbf{13}$$

19.
(13)

(a) $x + 3y = -3$

(a$'$) $x = -3y - 3$

(b) $3x - 5y = 19$

Substitute (a$'$) into (b) to get:

(b) $3(-3y - 3) - 5y = 19$

$$-14y - 9 = 19$$

$$-14y = 28$$

$$y = -2$$

(a$'$) $x = -3(-2) - 3 = 3$

$\mathbf{(3, -2)}$

20.
(14)

Graph the line to find the slope.

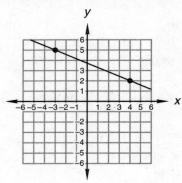

Slope $= -\dfrac{3}{7}$

$$y = -\frac{3}{7}x + b$$

$$5 = -\frac{3}{7}(-3) + b$$

$$\frac{35}{7} = \frac{9}{7} + b$$

$$\frac{26}{7} = b$$

$$y = -\frac{3}{7}x + \frac{26}{7}$$

Test 5

1.
(18)

$$\frac{20}{120} = \frac{A}{1560}$$

$$20 \cdot 1560 = 120A$$

$$31{,}200 = 120A$$

$$A = \mathbf{260\ liters}$$

2.
(5)

If $\dfrac{3}{10}$ missed, $\dfrac{7}{10}$ did not miss the demonstration.

$$F \times of = is$$

$$\frac{7}{10} \times N = 21$$

$$N = 21 \cdot \frac{10}{7} = 30$$

$M = 30 - 21 = \mathbf{9\ students}$

3.
(9)

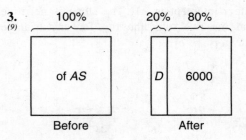

$$\frac{P}{100} \times of = is \quad \longrightarrow \quad \frac{80}{100} \times AS = 6000$$

$$AS = 6000 \cdot \frac{100}{80} = 7500$$

$$7500 - 6000 = \textbf{1500 spectators}$$

4.
(15)
(a) $2x + y = 16$

(b) $3x - 3y = 15$

3(a) $6x + 3y = 48$

(b) $\underline{3x - 3y = 15}$

$9x = 63$

$x = 7$

(a) $2(7) + y = 16$

$ y = 2$

(7, 2)

5.
(11)
$$\frac{a}{c^2} - a - \frac{2c}{3a^2} = \frac{3a^3}{3a^2 c^2} - \frac{3a^3 c^2}{3a^2 c^2} - \frac{2c^3}{3a^2 c^2}$$

$$= \frac{3a^3 - 3a^3 c^2 - 2c^3}{3a^2 c^2}$$

6.
(10)
$$c^2 = a^2 + b^2$$
$$14^2 = d^2 + 5^2$$
$$196 = d^2 + 25$$
$$171 = d^2$$
$$d = \sqrt{171} = \mathbf{3\sqrt{19}}$$

7.
(8)
(a) $y = -2$

(b) $y = -3x$

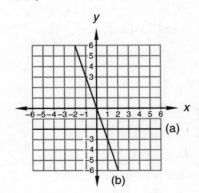

(a)

(b)

8.
(14)
$$y = -\frac{2}{3}x + b$$

$$-3 = -\frac{2}{3}(4) + b$$

$$-\frac{9}{3} = -\frac{8}{3} + b$$

$$-\frac{1}{3} = b$$

Since $m = -\frac{2}{3}$ and $b = -\frac{1}{3}$, $\mathbf{y = -\frac{2}{3}x - \frac{1}{3}}$.

9.
(17)
$$a + 65 + 90 = 180$$
$$a = \mathbf{25}$$

Since vertical angles are equal angles:

(a) $5x - 10y = 90$

(b) $-6x - 7y = 25$

6(a) $30x - 60y = 540$

5(b) $\underline{-30x - 35y = 125}$

$-95y = 665$

$y = \mathbf{-7}$

(a) $5x - 10(-7) = 90$

$5x = 20$

$x = \mathbf{4}$

10.
(B)
$$A = \frac{135}{360} \cdot \pi(10)^2$$

$$= \frac{135}{360} \cdot 100\pi \approx \mathbf{117.75\ cm^2}$$

11.
(4)
$$\frac{mn^0}{-n^2 n^{-2}}\left(\frac{m}{n^3} - \frac{2m^3 n^2}{mn^3}\right)$$

$$= \frac{mn^0 m}{-n^2 n^{-2} n^3} - \frac{mn^0 2m^3 n^2}{-n^2 n^{-2} mn^3}$$

$$= -\frac{m^2}{n^3} + \frac{2m^4 n^2}{mn^3} = \mathbf{-\frac{m^2}{n^3} + \frac{2m^3}{n}}$$

12.
(17)
(a) $R_A T_A + R_B T_B = 300$

(b) $R_A = 60$

(c) $R_B = 12$

(d) $T_B = T_A + 7$

Substitute (b), (c), and (d) into (a) to get:

$$60T_A + 12(T_A + 7) = 300$$

$$60T_A + 12T_A + 84 = 300$$

$$72T_A = 216$$

$$T_A = \mathbf{3}$$

(d) $T_B = 3 + 7 = \mathbf{10}$

13.
(16)
$$(3x - 2)(4x^2 - 7x - 5)$$

$$= 12x^3 - 21x^2 - 15x - 8x^2 + 14x + 10$$

$$= \mathbf{12x^3 - 29x^2 - x + 10}$$

14.
(16)

$$x - 3 \overline{\smash{)}\, x^3 - 2x^2 + 0x - 1} \quad \Big(x^2 + x + 3 + \tfrac{8}{x-3}\Big)$$

$$\underline{x^3 - 3x^2}$$
$$x^2 + 0x$$
$$\underline{x^2 - 3x}$$
$$3x - 1$$
$$\underline{3x - 9}$$
$$8$$

Check by multiplying:

$$(x - 3)\left(x^2 + x + 3 + \frac{8}{x - 3}\right)$$

$$= x^3 - 3x^2 + x^2 - 3x + 3x - 9 + 8$$

$$= x^3 - 2x^2 - 1$$

15.
(4)

$$-2[(-3)(-2 - x) + 2(x - 3)] = -2x$$
$$-12 - 6x - 4x + 12 = -2x$$
$$-10x = -2x$$
$$-8x = 0$$
$$x = \mathbf{0}$$

16.
(4)

$$3\frac{1}{4}x - \frac{4}{5} = -\frac{3}{20}$$

$$\frac{13}{4}x = -\frac{3}{20} + \frac{4}{5}$$

$$\frac{13}{4}x = -\frac{3}{20} + \frac{16}{20}$$

$$\frac{13}{4}x = \frac{13}{20}$$

$$x = \frac{13}{20} \cdot \frac{4}{13} = \mathbf{\frac{1}{5}}$$

17.
(2)

$$\frac{(-3a^0)^2 \, a^3 ac^2 c^0 c}{-3^{-3} \, a^2 cc^0 c^{-2} a} = \frac{-243 a^0 a^3 ac^2 c^0 c}{a^2 cc^0 c^{-2} a}$$

$$= \frac{-243 a^4 c^3}{a^3 c^{-1}} = \mathbf{-243 a c^4}$$

18.
(3)

$$\frac{4m^3}{n^2} + 2mn^{-2}m^2 - 6\frac{mm}{n^2}$$

$$= \frac{4m^3}{n^2} + \frac{2m^3}{n^2} - \frac{6m^2}{n^2} = \mathbf{\frac{6m^3 - 6m^2}{n^2}}$$

19.
(3)

$$x - xy(y^2 - x^2)$$

$$= -\frac{1}{3} - \left(-\frac{1}{3}\right)\left(\frac{1}{2}\right)\left[\left(\frac{1}{2}\right)^2 - \left(-\frac{1}{3}\right)^2\right]$$

$$= -\frac{1}{3} + \frac{1}{6}\left[\frac{1}{4} - \frac{1}{9}\right] = -\frac{1}{3} + \frac{1}{6}\left[\frac{5}{36}\right]$$

$$= \mathbf{-\frac{67}{216}}$$

20.
(A)

$$-2^2\big[(-6 + 4) - |-2 + 5| - 3(-3^0 + (-3))\big]$$

$$= -4[-2 - 3 - 3(-1 - 3)] = -4[7] = \mathbf{-28}$$

TEST 6

1.
(22)

$$D_M = D_K \text{ so } R_M T_M = R_K T_K$$

$$R_K = R_M - 10; \ T_M = 5; \ T_K = 6$$

$$R_M(5) = (R_M - 10)(6)$$

$$5R_M = 6R_M - 60$$

$$R_M = 60$$

$$D_M = 60(5) = \mathbf{300 \text{ miles}}$$

2.
(19)

(a) $N_E + N_N = 28$

(b) $45N_E + 90N_N = 1800$

Substitute $N_E = 28 - N_N$ into (b) to get:

(b′) $45(28 - N_N) + 90N_N = 1800$

$$45N_N = 540$$

$$N_N = \mathbf{12 \text{ notebooks}}$$

3.
(21)

(a) $\dfrac{N}{D} = \dfrac{8}{5} \quad\longrightarrow\quad 5N = 8D$

(b) $N - D = 24 \quad\longrightarrow\quad N = 24 + D$

Substitute $N = 24 + D$ into (a) to get:

(a′) $5(24 + D) = 8D$

$$120 + 5D = 8D$$

$$120 = 3D$$

$$\mathbf{40} = D$$

(b) $N = 24 + (40) = \mathbf{64}$

4.
(13)

(a) $3x + y = -1$

(a′) $y = -3x - 1$

(b) $4x + 3y = 12$

Substitute (a′) into (b) to get:

(b) $4x + 3(-3x - 1) = 12$

$$-5x - 3 = 12$$

$$-5x = 15$$

$$x = -3$$

(a′) $y = -3(-3) - 1 = 8$

$(\mathbf{-3, 8})$

5.
(16)

$$\frac{4x^2 + 4x + 4}{x-1\,\overline{)\,4x^3 - 0x^2 - 0x + 3}} + \frac{7}{x-1}$$

$$\underline{4x^3 - 4x^2}$$
$$\qquad 4x^2 - 0x$$
$$\qquad \underline{4x^2 - 4x}$$
$$\qquad\qquad 4x + 3$$
$$\qquad\qquad \underline{4x - 4}$$
$$\qquad\qquad\qquad 7$$

Check by multiplying:

$$\frac{4x^2(x-1)}{x-1} + \frac{4x(x-1)}{x-1} + \frac{4(x-1)}{x-1} + \frac{7}{x-1}$$

$$= \frac{4x^3 + 3}{x-1}$$

6.
(11)

$$\frac{5x^3}{z} + 2b^2 - \frac{3x^2}{b^3z} = \frac{5b^3x^3}{b^3z} + \frac{2b^5z}{b^3z} - \frac{3x^2}{b^3z}$$

$$= \frac{5b^3x^3 + 2b^5z - 3x^2}{b^3z}$$

7.
(12)

(a) Every point is 4 units to the left of the y-axis.

$$x = -4$$

(b) The y-intercept is 3. The slope is positive and the rise over the run for any triangle drawn is $\frac{1}{3}$.

$$y = \frac{1}{3}x + 3$$

8.
(22)

$$10 \times \overline{SF} = 4$$

$$\overline{SF} = \frac{2}{5}$$

$$a = 9\left(\frac{2}{5}\right) = \frac{18}{5}$$

$$b = 8\left(\frac{2}{5}\right) = \frac{16}{5}$$

9.
(20) Write the equation of the given line in slope-intercept form.

$$x - 2y = -7$$
$$-2y = -x - 7$$
$$y = \frac{1}{2}x + \frac{7}{2}$$

Since parallel lines have the same slope:

$$y = \frac{1}{2}x + b$$
$$-3 = \frac{1}{2}(-3) + b$$

$$-\frac{6}{2} + \frac{3}{2} = b$$

$$-\frac{3}{2} = b$$

$$y = \frac{1}{2}x - \frac{3}{2}$$

10.
(20)

$$5\sqrt{3}(2\sqrt{8} - 3\sqrt{3})$$
$$= 10\sqrt{2}\sqrt{2}\sqrt{2}\sqrt{3} - 15\sqrt{3}\sqrt{3} = \mathbf{20\sqrt{6} - 45}$$

11.
(20)

$$-6\sqrt{2}(5\sqrt{2} - 5\sqrt{10})$$
$$= -30\sqrt{2}\sqrt{2} + 30\sqrt{2}\sqrt{2}\sqrt{5} = \mathbf{-60 + 60\sqrt{5}}$$

12.
(21)

$$\frac{(0.008 \times 10^{-6})(300 \times 10^5)}{(0.006)(400,000,000)}$$

$$= \frac{(8 \times 10^{-9})(3 \times 10^7)}{(6 \times 10^{-3})(4 \times 10^8)} = \frac{24 \times 10^{-2}}{24 \times 10^5}$$

$$= \mathbf{1 \times 10^{-7}}$$

13.
(10)

$$m + 3 = 8$$
$$m = 5$$

$$m + n = 20$$
$$n = 20 - (5) = \mathbf{15}$$

14.
(6)

$$0.4(x - 3) = 0.04(3x + 40)$$
$$40(x - 3) = 4(3x + 40)$$
$$40x - 120 = 12x + 160$$
$$28x = 280$$
$$x = \mathbf{10}$$

15.
(4)

$$-3\left[x - (-4 - 3^0) - 2\right] + \left[(x + 3)(-2)\right] = 2x$$
$$-3x - 15 + 6 - 2x - 6 = 2x$$
$$-5x - 15 = 2x$$
$$-7x = 15$$
$$x = -\frac{15}{7}$$

16.
(4)

$$2x^{-2}y^3\left(\frac{x^2}{y^3} - 7x^{-2}y^{-3}\right)$$

$$= \frac{2x^{-2}y^3x^2}{y^3} - 14x^{-2}y^3x^{-2}y^{-3} = \mathbf{2 - 14x^{-4}}$$

17.
(2)

$$\frac{(a^2bc^{-1})^{-2}(ab^0c)^2}{(3a^2)^{-3}} = \frac{a^{-4}b^{-2}c^2a^2b^0c^2}{3^{-3}a^{-6}}$$

$$= \mathbf{27a^4b^{-2}c^4}$$

18.
(3)

$$-\frac{3xy^2}{x^{-1}} + 5xy^2x^{-1} - \frac{2x^3}{xy^{-2}}$$

$$= -3x^2y^2 + 5y^2 - 2x^2y^2 = \mathbf{-5x^2y^2 + 5y^2}$$

19.
(2)
$$\frac{-2}{-2^{-3}} - \frac{3}{-3^{-2}} = 2^4 + 3^3 = 16 + 27 = \mathbf{43}$$

20.
(3)
$$ac^2(a^2 - c) = -\frac{1}{3}\left(\frac{1}{2}\right)^2\left[\left(-\frac{1}{3}\right)^2 - \frac{1}{2}\right]$$

$$= -\frac{1}{3}\left(\frac{1}{4}\right)\left[\frac{1}{9} - \frac{1}{2}\right] = -\frac{1}{12}\left[-\frac{7}{18}\right] = \frac{\mathbf{7}}{\mathbf{216}}$$

TEST 7

1.
(22)

$D_B = D_C$ so $R_B T_B = R_C T_C$

$R_B = R_C + 10;\ T_B = 16;\ T_C = 20$

$(R_C + 10)16 = R_C(20)$

$16R_C + 160 = 20R_C$

$160 = 4R_C$

$40 = R_C$

$D_C = 40(20) = \mathbf{800\ miles}$

2.
(19)
(a) $N_D + N_Q = 168$

(b) $10N_D + 25N_Q = 3000$

-10(a) $-10N_D - 10N_Q = -1680$

(b) $\underline{\quad 10N_D + 25N_Q = \ \ 3000}$

$15N_Q = 1320$

$N_Q = \mathbf{88\ quarters}$

(a) $N_D + (88) = 168$

$N_D = \mathbf{80\ dimes}$

3.
(21)
(a) $2N_A = 3N_P - 11$

(b) $N_A + N_P = 12$

(b′) $N_A = 12 - N_P$

Substitute (b′) into (a) to get:

$2(12 - N_P) = 3N_P - 11$

$24 - 2N_P = 3N_P - 11$

$35 = 5N_P$

$\mathbf{7\ pears} = N_P$

4.
(9)

Before   After

$$\frac{P}{100} \times of = is \quad \longrightarrow \quad \frac{60}{100} \times N_S = 3600$$

$$N_S = 3600 \cdot \frac{100}{60} = \mathbf{6000\ signs}$$

5.
(16)
$$
\begin{array}{r}
4x^3 + 4x^2 + 4x + 4 + \frac{1}{x-1} \\
x - 1 \overline{)\ 4x^4 + 0x^3 + 0x^2 + 0x - 3} \\
\underline{4x^4 - 4x^3} \\
4x^3 + 0x^2 \\
\underline{4x^3 - 4x^2} \\
4x^2 + 0x \\
\underline{4x^2 - 4x} \\
4x - 3 \\
\underline{4x - 4} \\
1
\end{array}
$$

Check by multiplying:

$$(x - 1)\left(4x^3 + 4x^2 + 4x + 4 + \frac{1}{x-1}\right)$$

$$= 4x^4 - 4x^3 + 4x^3 - 4x^2 + 4x^2 - 4x + 4x$$
$$\quad - 4 + 1$$

$$= 4x^4 - 3$$

6.
(20)
$-3\sqrt{24} + 4\sqrt{81} + 3\sqrt{54}$

$= -3\sqrt{2}\sqrt{2}\sqrt{2}\sqrt{3} + 4\sqrt{3}\sqrt{3}\sqrt{3}\sqrt{3}$
$\quad + 3\sqrt{2}\sqrt{3}\sqrt{3}\sqrt{3}$

$= -6\sqrt{6} + 4 \cdot 9 + 9\sqrt{6} = \mathbf{3\sqrt{6} + 36}$

7.
(20)
$3\sqrt{6}(\sqrt{15} - 2\sqrt{6})$

$= 3\sqrt{2}\sqrt{3}\sqrt{3}\sqrt{5} - 6\sqrt{2}\sqrt{2}\sqrt{3}\sqrt{3}$

$= \mathbf{9\sqrt{10} - 36}$

8.
(25)
$$\frac{12x^2z - 3x}{3x} = \frac{3x(4xz - 1)}{3x} = \mathbf{4xz - 1}$$

9.
(21)
$$\frac{(0.0005 \times 10^{-3})(40 \times 10^4)}{(100{,}000)(0.2 \times 10^{-15})}$$

$$= \frac{(5 \times 10^{-7})(4 \times 10^5)}{(1 \times 10^5)(2 \times 10^{-16})} = \frac{20 \times 10^{-2}}{2 \times 10^{-11}}$$

$$= 10 \times 10^9 = \mathbf{1 \times 10^{10}}$$

10.
(11)
$$\frac{a^3c^2}{x} + x^3 - \frac{cx^2}{m} = \frac{a^3c^2m}{mx} + \frac{mx^4}{mx} - \frac{cx^3}{mx}$$

$$= \frac{a^3c^2m + mx^4 - cx^3}{mx}$$

11. (a) $8x + y = 4$
(23)
(a′) $y = -8x + 4$

(b) $4x - 12y = 12$

$\qquad 12y = 4x - 12$

$\qquad y = \frac{1}{3}x - 1$

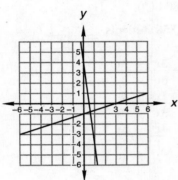

Substitute (a′) into (b) to get:

(b′) $4x - 12(-8x + 4) = 12$

$\qquad 4x + 96x - 48 = 12$

$\qquad 100x = 60$

$\qquad x = \frac{3}{5}$

(a′) $y = -8\left(\frac{3}{5}\right) + 4 = -\frac{4}{5}$

$\left(\dfrac{3}{5}, -\dfrac{4}{5}\right)$

12. Write the equation of the given line in slope-
(20) intercept form.

$2x - 6y = 7$

$\qquad y = \frac{1}{3}x - \frac{7}{6}$

Since parallel lines have the same slope:

$y = \frac{1}{3}x + b$

$8 = \frac{1}{3}(-3) + b$

$9 = b$

$y = \frac{1}{3}x + 9$

13. (a) $15x - 3y = 135$
(17)
(b) $-5x - 8y = 45$

$\qquad$ (a) $\quad 15x - 3y = 135$
3(b) $\quad \underline{-15x - 24y = 135}$
$\qquad\qquad\qquad -27y = 270$
$\qquad\qquad\qquad\qquad y = -10$

(a) $15x - 3(-10) = 135$

$\qquad\qquad 15x = 105$

$\qquad\qquad x = 7$

$10 \times \overrightarrow{SF} = 8$

$\qquad \overrightarrow{SF} = \frac{8}{10}$

$z \times \overrightarrow{SF} = 9$

$z \times \frac{8}{10} = 9$

$\qquad z = \frac{45}{4}$

14. $6^2 + 8^2 = B^2$
(26)
$\quad 36 + 64 = B^2$

$\qquad\quad 100 = B^2$

$\qquad\quad\; 10 = B$

$6 \cdot \overrightarrow{SF} = 9$

$\qquad \overrightarrow{SF} = \frac{3}{2}$

$8 \cdot \overrightarrow{SF} = A$

$8\left(\frac{3}{2}\right) = A$

$\qquad 12 = A$

$B \cdot \overrightarrow{SF} = B + C$

$10\left(\frac{3}{2}\right) = 10 + C$

$\qquad 15 = 10 + C$

$\qquad\;\; 5 = C$

15. $\dfrac{2x}{3} + \dfrac{3x - 4}{2} = 3$
(24)
$\qquad 4x + 9x - 12 = 18$

$\qquad\qquad 13x = 30$

$\qquad\qquad x = \frac{30}{13}$

16. $-4[x - (-2 - 3^0) - 2] + [(x - 2)(-3)] = 2x$
(4)
$$-4[x - (-3) - 2] + (-3x + 6) = 2x$$
$$-4(x + 1) + (-3x + 6) = 2x$$
$$-4x - 4 - 3x + 6 = 2x$$
$$2 = 9x$$
$$\frac{2}{9} = x$$

17. $\dfrac{3x}{4} - \dfrac{2x - 1}{5} = 3\dfrac{1}{2}$
(24)
$$15x - 8x + 4 = 70$$
$$7x = 66$$
$$x = 9\frac{3}{7}$$

18. $3ax^2 - 42a + 15ax = 3a(x^2 + 5x - 14)$
(26)
$$= \mathbf{3a(x + 7)(x - 2)}$$

19. $40cb - 2cbx^2 + 2bxc = 2bc(20 - x^2 + x)$
(26)
$$= -2bc(x^2 - x - 20) = \mathbf{-2bc(x - 5)(x + 4)}$$

20. $a^0 - ax^0(x^2 - a) - \dfrac{1}{-2^{-2}}$
(3)
$$= 1 - \left(-\frac{1}{2}\right)\left[(-3)^2 - \left(-\frac{1}{2}\right)\right] + 4$$
$$= 1 + \frac{1}{2}\left(9 + \frac{1}{2}\right) + 4 = 5 + \frac{19}{4} = \mathbf{\frac{39}{4}}$$

TEST 8

1.
(29)

$$R_K T_K + R_S T_S = 600$$
$$T_K = 6; \quad T_S = 4; \quad R_S = R_K + 5$$
$$6R_K + 4(R_K + 5) = 600$$
$$6R_K + 4R_K + 20 = 600$$
$$10R_K = 580$$
$$R_K = \mathbf{58 \ mph}$$

2. $7N \qquad 7N + 7 \qquad 7N + 14$
(6)
$$3(7N) = 2(7N + 14) + 14$$
$$21N = 14N + 28 + 14$$
$$7N = 42$$
$$N = 6$$

The desired multiples are **42, 49,** and **56.**

3. $\dfrac{P}{100} \times of = is$
(9)
$$\frac{260}{100} \times N_P = 117$$
$$N_P = 117 \cdot \frac{100}{260}$$
$$N_P = \mathbf{45 \ pansies}$$

4. $\dfrac{220}{5280} = \dfrac{869}{S}$
(18)
$$220S = 5280(869)$$
$$S = \mathbf{20{,}856 \ students}$$

5.
(16)
$$\begin{array}{r} 3x^2 + 3x + 3 - \frac{2}{x-1} \\ x - 1 \overline{)\ 3x^3 + 0x^2 + 0x - 5} \\ \underline{3x^3 - 3x^2} \\ 3x^2 + 0x \\ \underline{3x^2 - 3x} \\ 3x - 5 \\ \underline{3x - 3} \\ -2 \end{array}$$

Check by multiplying:
$$(x - 1)\left(3x^2 + 3x + 3 - \frac{2}{x - 1}\right)$$
$$= 3x^3 - 3x^2 + 3x^2 - 3x + 3x - 3 - 2$$
$$= 3x^3 - 5$$

6. $20xz - 2x^2z - 42z = 2z(10x - x^2 - 21)$
(26)
$$= -2z(x^2 - 10x + 21) = \mathbf{-2z(x - 7)(x - 3)}$$

7. $8xa + 42a - 2ax^2 = 2a(4x + 21 - x^2)$
(26)
$$= -2a(x^2 - 4x - 21) = \mathbf{-2a(x - 7)(x + 3)}$$

8. $63a^2xy + 9a^2x^2y + 108a^2y$
(26)
$$= 9a^2y(x^2 + 7x + 12) = \mathbf{9a^2y(x + 4)(x + 3)}$$

9. $\dfrac{pm - 3p^2m^3}{pm} = \dfrac{pm(1 - 3pm^2)}{pm} = \mathbf{1 - 3pm^2}$
(25)

10. $\dfrac{5}{2\sqrt{5}} \cdot \dfrac{\sqrt{5}}{\sqrt{5}} = \dfrac{5\sqrt{5}}{2(5)} = \mathbf{\dfrac{\sqrt{5}}{2}}$
(28)

11. $3\sqrt{27} - 4\sqrt{108} - 2\sqrt{75}$
(20)
$$= 3\sqrt{3}\sqrt{3}\sqrt{3} - 4\sqrt{2}\sqrt{2}\sqrt{3}\sqrt{3}\sqrt{3} - 2\sqrt{3}\sqrt{5}\sqrt{5}$$
$$= 9\sqrt{3} - 24\sqrt{3} - 10\sqrt{3} = \mathbf{-25\sqrt{3}}$$

12.
(28)
$$\frac{\dfrac{a + b}{x}}{\dfrac{c}{x}} \cdot \frac{\dfrac{x}{c}}{\dfrac{x}{c}} = \frac{a + b}{c}$$

13.
(27)
$$\frac{3x}{x^2 + 3x - 10} - \frac{4}{x + 5}$$

$$= \frac{3x}{(x + 5)(x - 2)} - \frac{4(x - 2)}{(x + 5)(x - 2)}$$

$$= \frac{3x - 4x + 8}{(x + 5)(x - 2)} = \frac{8 - x}{(x + 5)(x - 2)}$$

14.
(23)
(a) $3x + 2y = 12$

$$y = -\frac{3}{2}x + 6$$

(b) $8x - 2y = 10$

(b') $y = 4x - 5$

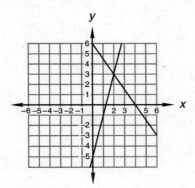

(a) $3x + 2y = 12$

(b) $\underline{8x - 2y = 10}$

$\quad 11x \qquad = 22$

$\qquad x = 2$

(b') $y = 4(2) - 5 = 3$

(2, 3)

15.
(12)
(a) The y-intercept is $+1$. The slope is negative and the rise over the run for any triangle drawn is $-\frac{3}{1}$.

$$y = -3x + 1$$

(b) Every point is 4 units below the x-axis.

$$y = -4$$

16.
(26)
$$\frac{6}{14} = \frac{M}{9}$$

$$14M = 54$$

$$M = \frac{27}{7}$$

$$N^2 = M^2 + 6^2$$

$$N^2 = \left(\frac{27}{7}\right)^2 + 36$$

$$N^2 = \frac{2493}{49}$$

$$N = \frac{\sqrt{9 \cdot 277}}{\sqrt{49}}$$

$$N = \frac{3\sqrt{277}}{7}$$

$$(N + P)^2 = 9^2 + 14^2$$

$$(N + P)^2 = 277$$

$$N + P = \sqrt{277}$$

$$\left(\frac{3\sqrt{277}}{7}\right) + P = \sqrt{277}$$

$$P = \frac{4\sqrt{277}}{7}$$

17.
(24)
$$\frac{3x}{4} - \frac{2x - 3}{3} = 6$$

$$9x - 8x + 12 = 72$$

$$x = 60$$

18.
(24)
$$-3x - \frac{5^0 - x}{2} + \frac{x + 4^0}{3} = 4$$

$$-18x - 3 + 3x + 2x + 2 = 24$$

$$-13x = 25$$

$$x = -\frac{25}{13}$$

19.
(3)
$$a^2 - b^3(a - b) = \left(\frac{1}{3}\right)^2 - \left(-\frac{1}{2}\right)^3\left(\frac{1}{3} + \frac{1}{2}\right)$$

$$= \frac{1}{9} + \frac{1}{8}\left(\frac{5}{6}\right) = \frac{16}{144} + \frac{15}{144} = \frac{31}{144}$$

20.
(B)
$$L.A. = \left[2\left(\frac{2\pi(4)}{2}\right) + 11.3\right] \times 8$$

$$= (8\pi + 11.3)8 \approx 291.36 \text{ m}^2$$

TEST 9

1.
(34)

$$R_C T_C = R_S T_S + 60$$
$$R_C = 60; \quad R_S = 40; \quad T_C = T_S$$
$$60T_C = 40T_C + 60$$
$$20T_C = 60$$
$$T_C = \textbf{3 hours}$$

2.
(6)

$$N \qquad N + 2 \qquad N + 4$$
$$3(N + N + 4) = 24 + 4(N + 2)$$
$$6N + 12 = 4N + 32$$
$$2N = 20$$
$$N = 10$$

The desired integers are **10, 12,** and **14.**

3.
(19)

(a) $N_N = N_D + 20$

(b) $5N_N + 10N_D = 700$

Substitute (a) into (b) to get:
$$5(N_D + 20) + 10N_D = 700$$
$$5N_D + 100 + 10N_D = 700$$
$$15N_D = 600$$
$$N_D = \textbf{40 dimes}$$

(a) $N_N = (40) + 20 = \textbf{60 nickles}$

4.
(9)

$$\frac{P}{100} \times of = is$$

$$\frac{80}{100} \times T = 680$$

$$T = 680 \cdot \frac{100}{80} = 850$$

$$M = 850 - 680 = \textbf{170 g}$$

5.
(31)

$$3x - 2y = 6$$
$$2y = 3x - 6$$
$$y = \frac{3}{2}x - 3$$

Since the slopes of perpendicular lines are negative reciprocals of each other:

$$y = -\frac{2}{3}x + b$$

$$2 = -\frac{2}{3}(1) + b$$

$$\frac{8}{3} = b$$

$$y = -\frac{2}{3}x + \frac{8}{3}$$

6.
(33)

$$\frac{\frac{a}{c} - 1}{\frac{1}{c} + 3} = \frac{\frac{a - c}{c}}{\frac{1 + 3c}{c}} \cdot \frac{\frac{c}{1 + 3c}}{\frac{c}{1 + 3c}} = \frac{a - c}{1 + 3c}$$

7.
(33)

$$\frac{\frac{3}{p} - 2x}{p} = \frac{\frac{3 - 2xp}{p}}{\frac{p}{1}} \cdot \frac{\frac{1}{p}}{\frac{1}{p}} = \frac{3 - 2xp}{p^2}$$

8.
(32)

$$-4\sqrt{\frac{3}{10}} + 2\sqrt{\frac{10}{3}}$$

$$= \frac{-4\sqrt{3}}{\sqrt{10}} \cdot \frac{\sqrt{10}}{\sqrt{10}} + \frac{2\sqrt{10}}{\sqrt{3}} \cdot \frac{\sqrt{3}}{\sqrt{3}}$$

$$= \frac{-4\sqrt{30}}{10} + \frac{2\sqrt{30}}{3} = \frac{-6\sqrt{30}}{15} + \frac{10\sqrt{30}}{15}$$

$$= \frac{4\sqrt{30}}{15}$$

9.
(21)

$$\frac{(50 \times 10^{-10})(0.0000024)}{8,000,000}$$

$$= \frac{(5 \times 10^{-9})(2.4 \times 10^{-6})}{8 \times 10^6} = \frac{1.2 \times 10^{-14}}{8 \times 10^6}$$

$$= \textbf{1.5} \times \textbf{10}^{-21}$$

10.
(20)

$$3\sqrt{63} - 5\sqrt{175} + \sqrt{252}$$
$$= 3\sqrt{3}\sqrt{3}\sqrt{7} - 5\sqrt{5}\sqrt{5}\sqrt{7} + \sqrt{2}\sqrt{2}\sqrt{3}\sqrt{3}\sqrt{7}$$
$$= 9\sqrt{7} - 25\sqrt{7} + 6\sqrt{7} = \textbf{--10}\sqrt{7}$$

11.
(25)

$$\frac{3a^2b^3 + ab^2}{ab^2} = \frac{ab^2(3ab + 1)}{ab^2} = \textbf{3ab + 1}$$

12.
(27)

$$\frac{2x}{x^2 + 4x - 12} - \frac{5}{x + 6}$$

$$= \frac{2x}{(x + 6)(x - 2)} - \frac{5(x - 2)}{(x + 6)(x - 2)}$$

$$= \frac{2x - 5x + 10}{(x + 6)(x - 2)} = \frac{\textbf{--3x + 10}}{\textbf{(x + 6)(x - 2)}}$$

13.
(16)

$$(a^2 + 2a + 5)(a^2 - a)$$
$$= a^4 + 2a^3 + 5a^2 - a^3 - 2a^2 - 5a$$
$$= \textbf{a}^4 + \textbf{a}^3 + \textbf{3a}^2 - \textbf{5a}$$

14.
(26)

$$-2x^2 + 14x + 16 = -2(x^2 - 7x - 8)$$
$$= \textbf{--2(x -- 8)(x + 1)}$$

15.
(26)

$$30x^2a - 3ax^3 - 75ax = -3ax(x^2 - 10x + 25)$$
$$= \textbf{--3ax(x -- 5)(x -- 5)}$$

16.
(24)

$$\frac{2x - 5}{3} - \frac{x}{6} = 12$$

$$4x - 10 - x = 72$$

$$3x = 82$$

$$x = \frac{82}{3}$$

17.
(4)

$$-3(x^0 - 2) + 4(-6^0 - x) = -2^0(2x - 5)$$

$$-3(1 - 2) + 4(-1 - x) = -1(2x - 5)$$

$$3 - 4 - 4x = -2x + 5$$

$$-2x = 6$$

$$x = -3$$

18. If x is the angle measure, $90 - x$ is the complement
(5) and $180 - x$ is the supplement.

$$2(180 - x) = (90 - x) + 250$$

$$360 - 2x = 340 - x$$

$$20° = x$$

19. (a) $6x + 4y = 2$
(23)

(a′) $y = -\dfrac{3}{2}x + \dfrac{1}{2}$

(b) $10x - 8y = 32$

$$y = \frac{5}{4}x - 4$$

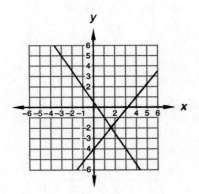

2(a) $12x + 8y = 4$

(b) $\underline{10x - 8y = 32}$

$$22x = 36$$

$$x = \frac{18}{11}$$

(a′) $y = -\dfrac{3}{2}\left(\dfrac{18}{11}\right) + \dfrac{1}{2} = -\dfrac{43}{22}$

$$\left(\frac{18}{11}, -\frac{43}{22}\right)$$

20.
(B)

$$A_{\text{Base}} = 24(6) + 12(6) + \frac{1}{2}\pi(6)^2$$

$$= 216 + 18\pi \approx 272.52 \text{ m}^2$$

$$V = A_{\text{Base}} \times H \approx (272.52)(8) \approx \mathbf{2180.16 \text{ m}^3}$$

$$P = 12 + 6 + 12 + 6 + 24 + \frac{1}{2} \cdot 2\pi(6)$$

$$= 60 + 6\pi \approx 78.84 \text{ m}$$

$$S.A. = 2A_{\text{Base}} + P \times H$$

$$\approx 2(272.52) + (78.84)(8) \approx \mathbf{1175.76 \text{ m}^2}$$

TEST 10

1.
(29)

$$R_S T_S + R_L T_L = 164$$

$$R_S = 6; \ R_L = 20; \ T_L = T_S + 3$$

$$6T_S + 20(T_S + 3) = 164$$

$$6T_S + 20T_S + 60 = 164$$

$$26T_S = 104$$

$$T_S = 4 \text{ hr}$$

Since Shanna started at noon, it was **4 p.m.** when she and Linda were 164 miles apart.

2. (a) $N_T = 2N_D + 3$
(19)

(b) $4N_T + 12N_D = 92$

Substitute (a) into (b) to get:

(b′) $4(2N_D + 3) + 12N_D = 92$

$$8N_D + 12 + 12N_D = 92$$

$$20N_D = 80$$

$$N_D = \mathbf{4 \text{ compact discs}}$$

(a) $N_T = 2(4) + 3 = \mathbf{11 \text{ audiocassette tapes}}$

3. Hydrogen: $2 \times 1 = 2$
(37)
Sulfur: $1 \times 32 = 32$

Oxygen: $4 \times 16 = 64$

Total: $= 98$

$$\frac{32}{98} = \frac{S}{1470}$$

$$98S = 32(1470)$$

$$S = \mathbf{480 \text{ grams}}$$

4. $f \times of = is$
(5)

$$2\frac{2}{3} \times T = 464$$

$$\frac{8}{3} \times T = 464$$

$$T = 464 \cdot \frac{3}{8}$$

$$T = \textbf{174 textbooks}$$

5. $(-8)^{-5/3} = \dfrac{1}{(-8)^{5/3}} = \dfrac{1}{((-8)^{1/3})^5} = \dfrac{1}{(-2)^5}$
(35)

$$= -\frac{1}{\textbf{32}}$$

6. $\dfrac{1}{-64^{-2/3}} = -64^{2/3} = -(64^{1/3})^2 = -4^2 = \textbf{-16}$
(35)

7. $3\sqrt{\dfrac{3}{13}} - 7\sqrt{\dfrac{13}{3}} = \dfrac{3\sqrt{3}}{\sqrt{13}} \cdot \dfrac{\sqrt{13}}{\sqrt{13}} - \dfrac{7\sqrt{13}}{\sqrt{3}} \cdot \dfrac{\sqrt{3}}{\sqrt{3}}$
(32)

$$= \frac{3\sqrt{39}}{13} - \frac{7\sqrt{39}}{3} = \frac{9\sqrt{39}}{39} - \frac{91\sqrt{39}}{39}$$

$$= -\frac{\textbf{82}\sqrt{\textbf{39}}}{\textbf{39}}$$

8. $3\sqrt{50}(2\sqrt{18} - \sqrt{6})$
(20)
$$= 3\sqrt{2}\sqrt{5}\sqrt{5}(2\sqrt{2}\sqrt{3}\sqrt{3} - \sqrt{2}\sqrt{3})$$
$$= \textbf{180} - \textbf{30}\sqrt{\textbf{3}}$$

9. $z + 70 + 2(60) = 360$
(35)
$$z + 190 = 360$$
$$z = \textbf{170}$$

$$2x = z + 70$$
$$2x = 240$$
$$x = \textbf{120}$$

$$y + 75 + 60 + 120 = 360$$
$$y + 255 = 360$$
$$y = \textbf{105}$$

10. Graph the line to find the slope.
(14)

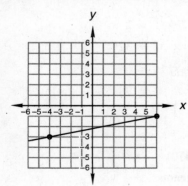

Slope $= \dfrac{+2}{+10} = \dfrac{1}{5}$

$$y = \frac{1}{5}x + b$$

$$-3 = \frac{1}{5}(-4) + b$$

$$-\frac{11}{5} = b$$

$$y = \frac{\textbf{1}}{\textbf{5}}x - \frac{\textbf{11}}{\textbf{5}}$$

11. $\dfrac{\dfrac{3ac}{x^2} + \dfrac{2}{z^2}}{1 - \dfrac{4}{x^2z^2}} = \dfrac{\dfrac{3acz^2 + 2x^2}{x^2z^2}}{\dfrac{x^2z^2 - 4}{x^2z^2}} \cdot \dfrac{\dfrac{x^2z^2}{x^2z^2 - 4}}{\dfrac{x^2z^2}{x^2z^2 - 4}}$
(33)

$$= \frac{\textbf{3}ac\textbf{z}^2 + \textbf{2}x^2}{x^2\textbf{z}^2 - \textbf{4}}$$

12. $\dfrac{x^2 - 5x - 6}{x^2 - 2x - 3} \div \dfrac{x^2 - 11x + 30}{x^2 - x - 20}$
(36)

$$= \frac{x^2 - 5x - 6}{x^2 - 2x - 3} \cdot \frac{x^2 - x - 20}{x^2 - 11x + 30}$$

$$= \frac{(x - 6)(x + 1)}{(x - 3)(x + 1)} \cdot \frac{(x - 5)(x + 4)}{(x - 5)(x - 6)} = \frac{x + 4}{x - 3}$$

13. $-2^2 - (-2)^2 - |2^2 - 3^2| - \dfrac{1}{-2^{-3}} - 3^0(-2^3 - 3^2)$
(2)

$$= -4 - 4 - |4 - 9| + 2^3 - 1(-8 - 9)$$

$$= -8 - 5 + 8 + 17 = \textbf{12}$$

14. $\dfrac{2}{x - 2} - \dfrac{2x - 1}{x^2 + x - 6} - \dfrac{2x}{x + 3}$
(27)

$$= \frac{2(x + 3)}{(x - 2)(x + 3)} - \frac{2x - 1}{(x - 2)(x + 3)}$$
$$- \frac{2x(x - 2)}{(x - 2)(x + 3)}$$

$$= \frac{2x + 6 - 2x + 1 - 2x^2 + 4x}{(x - 2)(x + 3)}$$

$$= \frac{-\textbf{2}x^2 + \textbf{4}x + \textbf{7}}{(x - \textbf{2})(x + \textbf{3})}$$

15. $\dfrac{5c^{-2}m^{-2}}{a^2}\left(\dfrac{c^2m}{a^{-2}} - \dfrac{2c^2m^2a^2}{t}\right) = \textbf{5}m^{-1} - \textbf{10}t^{-1}$
(4)

16. $(x + 3)^3 = (x + 3)(x + 3)(x + 3)$
(38)
$$= (x^2 + 6x + 9)(x + 3)$$
$$= x^3 + 6x^2 + 9x + 3x^2 + 18x + 27$$
$$= x^3 + \textbf{9}x^2 + \textbf{27}x + \textbf{27}$$

17.
(10)

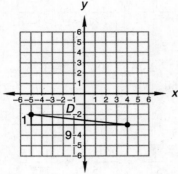

$$D^2 = 9^2 + 1^2$$
$$D^2 = 81 + 1$$
$$D^2 = 82$$
$$D = \sqrt{82}$$

18.
(24)
$$\frac{3 - 2x}{3} + \frac{x}{5} = 2$$
$$15 - 10x + 3x = 30$$
$$-7x = 15$$
$$x = -\frac{15}{7}$$

19.
(4)
$$3x(2 - 6^0) - 9x = 5x^0 - |-9 - 6| - 2(x - 4^0)$$
$$3x - 9x = 5 - 15 - 2x + 2$$
$$-4x = -8$$
$$x = 2$$

20.
(38)
$$-4x = 12 - x^2$$
$$x^2 - 4x - 12 = 0$$
$$(x + 2)(x - 6) = 0$$
$$x = -2, 6$$

TEST 11

1.
(21)
(a) $4N_L = 7N_S + 11$

(b) $N_S = N_L - 8$

Substitute (b) into (a) to get:

(a) $4N_L = 7(N_L - 8) + 11$
$$4N_L = 7N_L - 45$$
$$-3N_L = -45$$
$$N_L = 15$$

(b) $N_S = (15) - 8 = \mathbf{7}$

2.
(9)
$$\frac{P}{100} \times of = is$$
$$\frac{240}{100} \cdot 3640 = V$$
$$V = \textbf{8736 words}$$

3.
(37)
Potassium: $1 \times 39 = 39$
Chlorine: $1 \times 35 = 35$
Oxygen: $3 \times 16 = 48$
Total: $= 122$

$$\frac{35}{122} = \frac{Cl}{6832}$$
$$122Cl = 35(6832)$$
$$Cl = \textbf{1960 grams}$$

4.
(29)

$$R_A T_A + R_B T_B = 240$$
$$R_A = 20; R_B = 40; T_B = 7 - T_A$$
$$20T_A + 40(7 - T_A) = 240$$
$$-20T_A + 280 = 240$$
$$T_A = 2$$
$$R_A T_A = 20(2) = \textbf{40 miles}$$

5.
(39)
$$C = \frac{360 - 220}{2} = 70$$

Since diagonal bisectors of a rhombus are perpendicular, $A = 90$.

$$B + 90 + 70 = 180$$
$$B = \mathbf{20}$$

6.
(13)
$$6^2 = H^2 + \left(\frac{9}{2}\right)^2$$
$$36 = H^2 + \frac{81}{4}$$
$$\frac{63}{4} = H^2$$
$$\frac{3\sqrt{7}}{2} = H$$
$$A = \frac{9 \times \frac{3\sqrt{7}}{2}}{2} + \frac{320}{360}(\pi)(6)^2 \approx \textbf{118.34 in.}^2$$

7.
(22)
$$\frac{x}{4} = \frac{16}{5} \longrightarrow 5x = 64 \longrightarrow x = \frac{64}{5}$$
$$\frac{y}{3} = \frac{16}{5} \longrightarrow 5y = 48 \longrightarrow y = \frac{48}{5}$$

8.
(42)
$$\frac{(8351)(30,804 \times 10^{-9})}{(0.0061 \times 10^{20})(0.00079)}$$

$$\approx \frac{(8 \times 10^3)(3 \times 10^{-5})}{(6 \times 10^{17})(8 \times 10^{-4})} \approx \mathbf{5 \times 10^{-16}}$$

9.
(24)
$$\frac{-4(x + 2)}{3} - \frac{x}{5} = 9$$
$$-20x - 40 - 3x = 135$$
$$-23x = 175$$
$$x = -\frac{\mathbf{175}}{\mathbf{23}}$$

10.
(38)
$$15x^2 = 34x - x^3$$
$$x^3 + 15x^2 - 34x = 0$$
$$x(x + 17)(x - 2) = 0$$
$$x = \mathbf{0, -17, 2}$$

11.
(31)
$$2x + 2y = 18$$
$$2y = -2x + 18$$
$$y = -x + 9$$

Since the slopes of perpendicular lines are negative reciprocals of each other:
$$y = x + b$$
$$5 = -2 + b$$
$$7 = b$$
$$\mathbf{y = x + 7}$$

12.
(42)
$$\frac{by}{x} = c + \frac{a}{m}$$
$$mby = mcx + ax$$
$$mby - mcx = ax$$
$$m(by - cx) = ax$$
$$m = \frac{\mathbf{ax}}{\mathbf{by - cx}}$$

13.
(41)
$$20 \text{ mi}^3 \times \frac{5280 \text{ ft}}{1 \text{ mi}} \times \frac{5280 \text{ ft}}{1 \text{ mi}} \times \frac{5280 \text{ ft}}{1 \text{ mi}}$$
$$\times \frac{12 \text{ in.}}{1 \text{ ft}} \times \frac{12 \text{ in.}}{1 \text{ ft}} \times \frac{12 \text{ in.}}{1 \text{ ft}}$$
$$= \mathbf{20(5280)(5280)(5280)(12)(12)(12) \text{ in.}^3}$$

14.
(36)
$$\frac{x^3 - 2x^2 - 15x}{x^2 - 3x - 10} \div \frac{x^3 + 21x + 10x^2}{x^2 + 8x + 7}$$
$$= \frac{x^3 - 2x^2 - 15x}{x^2 - 3x - 10} \cdot \frac{x^2 + 8x + 7}{x^3 + 21x + 10x^2}$$
$$= \frac{x(x - 5)(x + 3)}{(x - 5)(x + 2)} \cdot \frac{(x + 7)(x + 1)}{x(x + 7)(x + 1)}$$
$$= \frac{\mathbf{x + 1}}{\mathbf{x + 2}}$$

15.
(35)
$$-27^{-4/3} = -\frac{1}{27^{4/3}} = -\frac{1}{(27^{1/3})^4} = -\frac{\mathbf{1}}{\mathbf{81}}$$

16.
(20)
$$2\sqrt{6}(3\sqrt{8} - 5\sqrt{6})$$
$$= 6\sqrt{2}\sqrt{2}\sqrt{2}\sqrt{2}\sqrt{3} - 10\sqrt{2}\sqrt{2}\sqrt{3}\sqrt{3}$$
$$= \mathbf{24\sqrt{3} - 60}$$

17.
(32)
$$4\sqrt{\frac{5}{7}} - 3\sqrt{\frac{7}{5}} = \frac{4\sqrt{5}}{\sqrt{7}} \cdot \frac{\sqrt{7}}{\sqrt{7}} - \frac{3\sqrt{7}}{\sqrt{5}} \cdot \frac{\sqrt{5}}{\sqrt{5}}$$
$$= \frac{4\sqrt{35}}{7} - \frac{3\sqrt{35}}{5} = \frac{20\sqrt{35}}{35} - \frac{21\sqrt{35}}{35}$$
$$= -\frac{\mathbf{\sqrt{35}}}{\mathbf{35}}$$

18.
(25)
$$\frac{6a^3 - 6a^6}{6a^2} = \frac{6a^3(1 - a^3)}{6a^2} = a(1 - a^3)$$
$$= \mathbf{a - a^4}$$

19.
(33)
$$\frac{\frac{a}{c} + 2}{5 - \frac{1}{c}} = \frac{\frac{a + 2c}{c}}{\frac{5c - 1}{c}} \cdot \frac{\frac{c}{5c - 1}}{\frac{c}{5c - 1}} = \frac{\mathbf{a + 2c}}{\mathbf{5c - 1}}$$

20.
(3)
$$-3x^{-1} - x^{-3} + x^{-2}$$
$$= -3\left(-\frac{1}{4}\right)^{-1} - \left(-\frac{1}{4}\right)^{-3} + \left(-\frac{1}{4}\right)^{-2}$$
$$= 12 + 64 + 16 = \mathbf{92}$$

TEST 12

1.
(21)
(a) $\dfrac{N_C}{N_D} = \dfrac{10}{3} \quad\longrightarrow\quad 3N_C = 10N_D$

(b) $N_C = 5N_D - 10$

Substitute (b) into (a) to get:

(a) $3(5N_D - 10) = 10N_D$
$$15N_D - 30 = 10N_D$$
$$5N_D = 30$$
$$N_D = 6$$

(b) $N_C = 5(6) - 10 = \mathbf{20 \text{ cassettes}}$

2.
(19)
(a) $N_N + N_Q = 144$

(a′) $N_Q = 144 - N_N$

(b) $5N_N + 25N_Q = 2000$

Substitute (a′) into (b) to get:

(b) $5N_N + 25(144 - N_N) = 2000$
$$-20N_N + 3600 = 2000$$
$$-20N_N = -1600$$
$$N_N = \mathbf{80 \text{ nickels}}$$

(a′) $N_Q = 144 - 80 = \mathbf{64 \text{ quarters}}$

3. $\frac{P}{100} \times of = is$
$^{(9)}$

$$\frac{180}{100} \times S = 7200$$

$$S = 7200 \cdot \frac{100}{180} = \textbf{4000}$$

4.
$^{(34)}$

$$D_B$$
$$\xrightarrow{\hspace{3cm}}$$
$$\xrightarrow{D_J} \quad 65$$

$R_B T_B - R_J T_J = 65$

$R_B = 60;\ R_J = 50;\ T_B = T_J$

$60T_B - 50T_B = 65$

$10T_B = 65$

$T_B = \textbf{6.5 hours}$

5. $\tan A = \frac{5}{12}$
$^{(44)}$

$A \approx \textbf{22.62}$

$B + 22.62 + 90 \approx 180$

$B \approx \textbf{67.38}$

$\sin 22.62° \approx \frac{5}{m}$

$m \approx \frac{5}{\sin 22.62°} \approx \textbf{13}$

6. $F + 32 + 90 = 180$
$^{(44)}$

$F = \textbf{58}$

$\tan 32° = \frac{m}{5}$

$m = 5\tan 32° \approx \textbf{3.12}$

$\cos 32° = \frac{5}{n}$

$n = \frac{5}{\cos 32°} \approx \textbf{5.90}$

7. $\frac{2c}{a} - b = \frac{m}{x}$
$^{(40)}$

$2cx - bax = ma$

$2cx = ma + bax$

$2cx = a(m + bx)$

$\frac{2cx}{m + bx} = a$

8.
$^{(23)}$

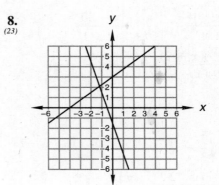

(a) $3x - 4y = -12$

(a′) $y = \frac{3}{4}x + 3$

(b) $6x + 2y = -3$

(b′) $y = -3x - \frac{3}{2}$

Substitute (a′) into (b′) to get:

(b′) $-3x - \frac{3}{2} = \frac{3}{4}x + 3$

$-\frac{15}{4}x = \frac{9}{2}$

$\phantom{(b') -\frac{15}{4}}x = -\frac{6}{5}$

(a′) $y = \frac{3}{4}\left(-\frac{6}{5}\right) + 3 = \frac{21}{10}$

$$\left(-\frac{6}{5}, \frac{21}{10}\right)$$

9. $\left(x - \frac{2}{3}\right)^2 = 14$
$^{(45)}$

$x - \frac{2}{3} = \pm\sqrt{14}$

$x = \frac{2}{3} \pm \sqrt{14}$

10. $\frac{-2x + 3}{4} - \frac{3x + 5}{3} = 5$
$^{(24)}$

$-6x + 9 - 12x - 20 = 60$

$-18x = 71$

$x = -\frac{71}{18}$

11. $20x = x^3 - x^2$
$^{(38)}$

$0 = x^3 - x^2 - 20x$

$0 = x(x - 5)(x + 4)$

$x = \textbf{0, 5, -4}$

12. $\dfrac{x^3 - 7x^2 - 18x}{x^2 - 81} \cdot \dfrac{x^2 + x - 30}{x^3 + 8x^2 + 12}$
(36)

$= \dfrac{x(x - 9)(x + 2)}{(x + 9)(x - 9)} \cdot \dfrac{(x + 6)(x - 5)}{x(x + 6)(x + 2)}$

$= \dfrac{x - 5}{x + 9}$

13. $3\sqrt{\dfrac{3}{7}} - 2\sqrt{\dfrac{7}{3}} - \sqrt{189}$
(46)

$= \dfrac{3\sqrt{3}}{\sqrt{7}} \cdot \dfrac{\sqrt{7}}{\sqrt{7}} - \dfrac{2\sqrt{7}}{\sqrt{3}} \cdot \dfrac{\sqrt{3}}{\sqrt{3}} - 3\sqrt{21}$

$= \dfrac{3\sqrt{21}}{7} - \dfrac{2\sqrt{21}}{3} - 3\sqrt{21}$

$= \dfrac{9\sqrt{21}}{21} - \dfrac{14\sqrt{21}}{21} - \dfrac{63\sqrt{21}}{21}$

$= -\dfrac{68\sqrt{21}}{21}$

14. $\sqrt[4]{mn^5}\sqrt[3]{m^5n^3} = (mn^5)^{1/4}(m^5n^3)^{1/3}$
(46)

$= m^{1/4}n^{5/4}m^{5/3}n^1 = \boldsymbol{m^{23/12}n^{9/4}}$

15. $\sqrt{8\sqrt{8}} = \left[2^3(2^{3/2})\right]^{1/2} = (2^{9/2})^{1/2} = \boldsymbol{2^{9/4}}$
(46)

16. $\dfrac{c - \dfrac{3x^2}{c}}{2c + \dfrac{z^2}{c}} = \dfrac{\dfrac{c^2 - 3x^2}{c}}{\dfrac{2c^2 + z^2}{c}} \cdot \dfrac{\dfrac{c}{2c^2 + z^2}}{\dfrac{c}{2c^2 + z^2}}$
(33)

$= \dfrac{c^2 - 3x^2}{2c^2 + z^2}$

17. $\dfrac{1}{-16^{-3/4}} = -16^{3/4} = -(16^{1/4})^3 = -2^3 = \boldsymbol{-8}$
(35)

18. $\dfrac{(677{,}123)(5{,}134{,}692 \times 10^6)}{6023 \times 10^{-8}}$
(42)

$\approx \dfrac{(7 \times 10^5)(5 \times 10^{12})}{6 \times 10^{-5}} \approx \boldsymbol{6 \times 10^{22}}$

19. Graph the line to find the slope.
(31)

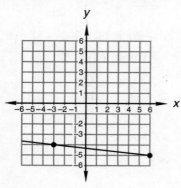

$m = \dfrac{-1}{+9} = -\dfrac{1}{9}$

Since the slopes of perpendicular lines are negative reciprocals of each other,

$m\perp = 9.$

$y = 9x + b$

$-5 = 9(3) + b$

$-32 = b$

$\boldsymbol{y = 9x - 32}$

20. $a^3 + ab^2 - ab = \left(-\dfrac{1}{3}\right)^3 - \dfrac{1}{3}\left(-\dfrac{1}{2}\right)^2 + \dfrac{1}{3}\left(-\dfrac{1}{2}\right)$
(3)

$= -\dfrac{1}{27} - \dfrac{1}{3}\left(\dfrac{1}{4}\right) - \dfrac{1}{6} = -\dfrac{4}{108} - \dfrac{9}{108} - \dfrac{18}{108}$

$= -\dfrac{31}{108}$

TEST 13

1. $N \qquad N + 2 \qquad N + 4$
(6)

$6(N + N + 2) = 8(N + 4) + 16$

$12N + 12 = 8N + 48$

$4N = 36$

$N = 9$

The desired integers are **9, 11,** and **13.**

2. Phosphorus: $1 \times 31 = 31$
(37)

Hydrogen: $\quad 3 \times 1 = 3$

Total: $\qquad\qquad = 34$

$\dfrac{31}{34} = \dfrac{P}{2720}$

$34P = 31(2720)$

$P = \boldsymbol{2480\ grams}$

3.
(29)

$\overset{D_M}{\longleftarrow}\ \overset{D_A}{\longrightarrow}$

$\underset{560}{\rule{3cm}{0.4pt}}$

$R_MT_M + R_AT_A = 560$

$R_M = 20;\ R_A = 50;\ T_M = T_A$

$20T_M + 50T_M = 560$

$70T_M = 560$

$T_M = 8$

$R_MT_M = 20(8) = \boldsymbol{160\ miles}$

4. $\dfrac{7\ mi}{hr} \times \dfrac{5280\ ft}{1\ mi} \times \dfrac{1\ hr}{60\ min} \times \dfrac{1\ min}{60\ s}$
(47)

$= \dfrac{7(5280)}{(60)(60)}\ \dfrac{ft}{s}$

5. $y = -4x + b$
(49)

Use the point $(4, 0)$ for x and y.

$0 = -4(4) + b$

$16 = b$

$\mathbf{y = -4x + 16}$

6. $B = 180 - 40 - 90 = \mathbf{50}$
(49)

$\sin 50° = \dfrac{a}{12}$

$a = 12 \sin 50° \approx \mathbf{9.19}$

7. $\dfrac{2x}{m} + \dfrac{3a}{x} = \dfrac{5}{c}$
(40)

$2cx^2 + 3acm = 5mx$

$2cx^2 = 5mx - 3acm$

$2cx^2 = m(5x - 3ac)$

$\dfrac{2cx^2}{5x - 3ac} = m$

8. $x^2 = 8 - 9x$
(50)

$\left(x^2 + 9x + \phantom{\dfrac{81}{4}}\right) = 8$

$x^2 + 9x + \dfrac{81}{4} = 8 + \dfrac{81}{4}$

$\left(x + \dfrac{9}{2}\right)^2 = \dfrac{113}{4}$

$x + \dfrac{9}{2} = \pm\dfrac{\sqrt{113}}{2}$

$\mathbf{x = -\dfrac{9}{2} \pm \dfrac{\sqrt{113}}{2}}$

9. $189x = 3x^3 + 6x^2$
(38)

$0 = 3x^3 + 6x^2 - 189x$

$0 = 3x(x + 9)(x - 7)$

$\mathbf{x = 0, -9, 7}$

10. $9 = 3 + \sqrt{x - 2}$
(48)

$6 = \sqrt{x - 2}$

$36 = x - 2$

$\mathbf{38 = x}$

11. $3x^0 - 4x(-1 - 3^0) = 5(x + 3)$
(4)

$3 + 8x = 5x + 15$

$3x = 12$

$\mathbf{x = 4}$

12. $\sqrt[3]{27\sqrt{3}} = \sqrt[3]{3^3\sqrt{3}} = [3^3(3^{1/2})]^{1/3} = 3(3^{1/6})$
(47)

$= \mathbf{3^{7/6}}$

13. $\sqrt[3]{a^2c^4}\sqrt{a^5c^3} = (a^2c^4)^{1/3}(a^5c^3)^{1/2}$
(46)

$= a^{2/3}c^{4/3}a^{5/2}c^{3/2} = \mathbf{a^{19/6}c^{17/6}}$

14. $6\sqrt{\dfrac{2}{3}} - 8\sqrt{\dfrac{3}{2}} - \sqrt{294}$
(46)

$= \dfrac{6\sqrt{2}}{\sqrt{3}} \cdot \dfrac{\sqrt{3}}{\sqrt{3}} - \dfrac{8\sqrt{3}}{\sqrt{2}} \cdot \dfrac{\sqrt{2}}{\sqrt{2}} - 7\sqrt{6}$

$= \dfrac{6\sqrt{6}}{3} - \dfrac{8\sqrt{6}}{2} - 7\sqrt{6}$

$= 2\sqrt{6} - 4\sqrt{6} - 7\sqrt{6} = \mathbf{-9\sqrt{6}}$

15. $\dfrac{x^3 - 6x + x^2}{-6 - 5x + x^2} \div \dfrac{2x^2 - 8x + x^3}{x^2 - 2x - 24}$
(36)

$= \dfrac{x^3 - 6x + x^2}{-6 - 5x + x^2} \cdot \dfrac{x^2 - 2x - 24}{2x^2 - 8x + x^3}$

$= \dfrac{x(x + 3)(x - 2)}{(x - 6)(x + 1)} \cdot \dfrac{(x - 6)(x + 4)}{x(x + 4)(x - 2)}$

$= \dfrac{\mathbf{x + 3}}{\mathbf{x + 1}}$

16.
(10)

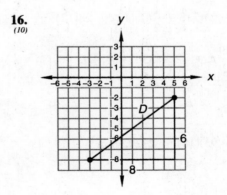

$D^2 = 8^2 + 6^2 = 64 + 36 = 100$

$D = \sqrt{100} = \mathbf{10}$

17. $\dfrac{(0.023 \times 10^{-8})(78,493 \times 10^5)}{18,000,000}$
(42)

$\approx \dfrac{(2 \times 10^{-10})(8 \times 10^9)}{2 \times 10^7} \approx \mathbf{8 \times 10^{-8}}$

18.
(26)

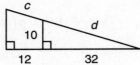

$$d^2 = 32^2 + 10^2 = 1024 + 100 = 1124$$
$$d = \sqrt{1124} = 2\sqrt{281}$$

$$\frac{32}{44} = \frac{2\sqrt{281}}{c + 2\sqrt{281}}$$
$$32c + 64\sqrt{281} = 88\sqrt{281}$$
$$32c = 24\sqrt{281}$$
$$c = \frac{3\sqrt{281}}{4}$$

19.
(1)
$$v + 60 + 90 = 180$$
$$v = \mathbf{30}$$
$$w + 60 + 90 = 180$$
$$w = \mathbf{30}$$
$$x + 60 + 90 = 180$$
$$x = \mathbf{30}$$
$$y + 30 + 90 = 180$$
$$y = \mathbf{60}$$

Since vertical angles are equal angles,
$$z = y = \mathbf{60}.$$

20.
(B)
$$A_{\text{Base}} = \frac{1}{2}(\pi)(7)^2 + \frac{1}{2}(7)(12) + 28(12)$$
$$= \frac{49}{2}\pi + 42 + 336 \approx 454.93 \text{ m}^2$$

$$V = \frac{1}{3}A_{\text{Base}} \times H$$
$$\approx \frac{1}{3}(454.93)(18) \approx \mathbf{2729.58 \text{ m}^3}$$

TEST 14

1.
(52)
Iodine P_N + Iodine D_N = Iodine Total
$$0.12(P_N) + 0.37(D_N) = 0.16(25)$$

(a) $0.12P_N + 0.37D_N = 4$

(b) $P_N + D_N = 25$

(b′) $D_N = 25 - P_N$

Substitute (b′) into (a) to get:

(a′) $0.12P_N + 0.37(25 - P_N) = 4$
$$0.12P_N + 9.25 - 0.37P_N = 4$$
$$-0.25P_N = -5.25$$
$$P_N = 21$$

(b′) $D_N = 25 - (21) = 4$

21 liters 12%, 4 liters 37%

2.
(53)
Sodium: $2 \times 23 = 46$

Sulfur: $2 \times 32 = 64$

Oxygen: $3 \times 16 = 48$

Total: $ = 158$

Sodium $= \dfrac{46}{158} \approx 0.29 \approx \mathbf{29\%}$

3.
(29)

$$\underset{660}{\overset{\overset{\displaystyle D_C \qquad D_B}{\longleftrightarrow}}{}}$$

$$R_C T_C + R_B T_B = 660$$
$$R_C = 30; \ R_B = 60;$$
$$T_C + T_B = 14$$
$$T_C = 14 - T_B$$
$$30(14 - T_B) + 60T_B = 660$$
$$420 - 30T_B + 60T_B = 660$$
$$30T_B = 240$$
$$T_B = 8$$

$$D_B = 60(8) = \mathbf{480 \text{ miles}}$$

4.
(51)
$$5i^5 - 3 + 2i - \sqrt{-36}$$
$$= 5i(ii)(ii) - 3 + 2i - 6i$$
$$= 5i - 3 - 4i = \mathbf{-3 + i}$$

5.
(51)
$$3i^3 + 5i - 3i^2 + \sqrt{-25} - 2$$
$$= 3i(ii) + 5i - 3(ii) + 5i - 2$$
$$= -3i + 5i + 3 + 5i - 2 = \mathbf{1 + 7i}$$

6.
(47)
$$\sqrt[3]{25\sqrt{5}} = \left[5^2(5^{1/2})\right]^{1/3} = 5^{2/3}5^{1/6} = \mathbf{5^{5/6}}$$

7.
(46)
$$\sqrt[5]{m^2 p^7}\sqrt[3]{m^5 p} = (m^2 p^7)^{1/5}(m^5 p)^{1/3}$$
$$= m^{2/5}p^{7/5}m^{5/3}p^{1/3} = \mathbf{m^{31/15}p^{26/15}}$$

8.
(46)
$$2\sqrt{\frac{7}{13}} - 3\sqrt{\frac{13}{7}} - \sqrt{364}$$

$$= \frac{2\sqrt{7}}{\sqrt{13}} \cdot \frac{\sqrt{13}}{\sqrt{13}} - \frac{3\sqrt{13}}{\sqrt{7}} \cdot \frac{\sqrt{7}}{\sqrt{7}} - 2\sqrt{91}$$

$$= \frac{2\sqrt{91}}{13} - \frac{3\sqrt{91}}{7} - 2\sqrt{91}$$

$$= \frac{14\sqrt{91}}{91} - \frac{39\sqrt{91}}{91} - \frac{182\sqrt{91}}{91} = -\frac{\mathbf{207}\sqrt{\mathbf{91}}}{\mathbf{91}}$$

9.
(35)
$$\frac{1}{-27^{-2/3}} = -27^{2/3} = -(27^{1/3})^2 = -3^2 = \mathbf{-9}$$

10.
(33)
$$\frac{\dfrac{r^2 c}{m^3} - c^2}{\dfrac{4}{m^3} - \dfrac{r}{m^3}} = \frac{\dfrac{r^2 c - c^2 m^3}{m^3}}{\dfrac{4-r}{m^3}} \cdot \frac{\dfrac{m^3}{4-r}}{\dfrac{m^3}{4-r}}$$

$$= \frac{\mathbf{r^2 c - c^2 m^3}}{\mathbf{4 - r}}$$

11.
(50)
$$-7x = 5 - x^2$$

$$\left(x^2 - 7x + \phantom{\frac{49}{4}}\right) = 5$$

$$x^2 - 7x + \frac{49}{4} = 5 + \frac{49}{4}$$

$$\left(x - \frac{7}{2}\right)^2 = \frac{69}{4}$$

$$x - \frac{7}{2} = \pm\frac{\sqrt{69}}{2}$$

$$x = \frac{\mathbf{7}}{\mathbf{2}} \pm \frac{\sqrt{\mathbf{69}}}{\mathbf{2}}$$

12.
(24)
$$\frac{5x - 3}{5} - \frac{4x + 2}{6} = 4$$

$$30x - 18 - 20x - 10 = 120$$

$$10x = 148$$

$$x = \frac{\mathbf{74}}{\mathbf{5}}$$

13.
(48)
$$\sqrt{x^2 - 3x - 6} - 3 = x - 2$$

$$\sqrt{x^2 - 3x - 6} = x + 1$$

$$x^2 - 3x - 6 = x^2 + 2x + 1$$

$$-5x = 7$$

$$x = -\frac{7}{5}$$

Check:
$$\sqrt{\frac{49}{25} + \frac{21}{5} - 6} - 3 = -\frac{7}{5} - 2$$

$$\sqrt{\frac{4}{25}} - 3 = -\frac{7}{5} - \frac{10}{5}$$

$$\frac{2}{5} - \frac{15}{5} = -\frac{17}{5}$$

$$-\frac{13}{5} = -\frac{17}{5} \qquad \text{not true}$$

No real number solution

14.
(20)
Graph the line to find the slope.

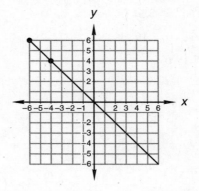

Slope $= \dfrac{-2}{+2} = -1$

Since the slopes of parallel lines are equal:

$$y = -x + b$$

$$-7 = -2 + b$$

$$-5 = b$$

$$\boldsymbol{y = -x - 5}$$

15.
(49)
$$x = 180 - 135 = 45$$

$$y = x = 45$$

$$\sin 45° = \frac{8}{a}$$

$$a = \frac{8}{\sin 45°} \approx \mathbf{11.31}$$

16.
(53)
$$400 \text{ km}^2 \times \frac{1000 \text{ m}}{1 \text{ km}} \times \frac{1000 \text{ m}}{1 \text{ km}} \times \frac{100 \text{ cm}}{1 \text{ m}}$$

$$\times \frac{100 \text{ cm}}{1 \text{ m}} \times \frac{1 \text{ in.}}{2.54 \text{ cm}} \times \frac{1 \text{ in.}}{2.54 \text{ cm}} \times \frac{1 \text{ ft}}{12 \text{ in.}}$$

$$\times \frac{1 \text{ ft}}{12 \text{ in.}} \times \frac{1 \text{ mi}}{5280 \text{ ft}} \times \frac{1 \text{ mi}}{5280 \text{ ft}}$$

$$= \frac{\mathbf{400(1000)(1000)(100)(100)}}{\mathbf{(2.54)(2.54)(12)(12)(5280)(5280)}} \text{ mi}^2$$

17.
(40)
$$\frac{m}{c} - 3a + \frac{e}{x} = f$$
$$mx - 3acx + ec = fcx$$
$$(m - 3ac - fc)x = -ec$$
$$x = \frac{ec}{3ac + fc - m}$$

18.
(54)

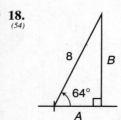

$$A = 8 \cos 64° \approx 3.51$$
$$B = 8 \sin 64° \approx 7.19$$
3.51 R + 7.19 U

19.
(54)
$$\frac{x}{a} = \frac{y}{b}$$
$$x = \frac{ay}{b}$$

20.
(11)
$$2n = 46$$
$$n = \mathbf{23}$$
$$q = n = \mathbf{23}$$
$$m + 26 + 23 = 180$$
$$m = \mathbf{131}$$
$$23 + 131 + p = 180$$
$$p = \mathbf{26}$$

TEST 15

1.
(57)
$$\frac{P_1 V_1}{T_1} = \frac{P_2 V_2}{T_2}$$
$$\frac{100(16)}{500} = \frac{P_2(20)}{400}$$
$$P_2 = \mathbf{64 \ N/m^2}$$

2.
(55)
$$N \quad N + 2 \quad N + 4$$
$$(N + 2)(N + 4) = 30N - 55$$
$$N^2 + 6N + 8 = 30N - 55$$
$$N^2 - 24N + 63 = 0$$
$$(N - 3)(N - 21) = 0$$
$$N = 3, 21$$

The desired integers are **3, 5,** and **7** or **21, 23,** and **25.**

3.
(52)
$$\text{Salt } P_N + \text{Salt } D_N = \text{Salt Total}$$
$$0.25(P_N) + 0.1(D_N) = 0.15(120)$$

(a) $0.25P_N + 0.1D_N = 18$

(b) $P_N + D_N = 120$

(b′) $D_N = 120 - P_N$

Substitute (b′) into (a) to get:

(a′) $0.25P_N + 0.1(120 - P_N) = 18$
$$0.15P_N + 12 = 18$$
$$0.15P_N = 6$$
$$P_N = \mathbf{40 \ liters \ of}$$
$$\mathbf{25\% \ solution}$$

(b′) $D_N = 120 - 40$
$$= \mathbf{80 \ liters \ of \ 10\% \ solution}$$

4.
(37)
Carbon: $6 \times 12 = 72$
Chlorine: $1 \times 35 = 35$
$$\frac{35}{72} = \frac{\text{Cl}}{936}$$
$$72\text{Cl} = 35(936)$$
$$\text{Cl} = \mathbf{455 \ grams}$$

5.
(55)
$$\frac{x - k}{p} - m = \frac{f}{a}$$
$$a(x - k - mp) = fp$$
$$a = \frac{fp}{x - k - mp}$$

6.
(54)

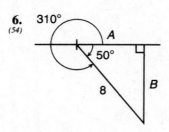

$$A = 8 \cos 50° \approx 5.14$$
$$B = 8 \sin 50° \approx 6.13$$
5.14 R − 6.13 U

7.
(47)
$$\frac{30 \text{ ft}}{\text{s}} \times \frac{12 \text{ in.}}{1 \text{ ft}} \times \frac{2.54 \text{ cm}}{1 \text{ in.}} \times \frac{1 \text{ m}}{100 \text{ cm}}$$
$$\times \frac{60 \text{ s}}{1 \text{ min}} \times \frac{60 \text{ min}}{1 \text{ hr}} = \frac{30(12)(2.54)(60)(60)}{100} \frac{\text{m}}{\text{hr}}$$

8.
(58)

$$3x^2 + 2x = 10$$

$$\left(x^2 + \frac{2}{3}x + \quad\right) = \frac{10}{3}$$

$$x^2 + \frac{2}{3}x + \frac{1}{9} = \frac{10}{3} + \frac{1}{9}$$

$$\left(x + \frac{1}{3}\right)^2 = \frac{31}{9}$$

$$x + \frac{1}{3} = \pm\frac{\sqrt{31}}{3}$$

$$x = -\frac{1}{3} \pm \frac{\sqrt{31}}{3}$$

9.
(48)

$$\sqrt{x^2 - 8x + 3} + 6 = x$$

$$x^2 - 8x + 3 = x^2 - 12x + 36$$

$$4x = 33$$

$$x = \frac{33}{4}$$

Check: $\sqrt{\left(\frac{33}{4}\right)^2 - 8\left(\frac{33}{4}\right) + 3} + 6 = \frac{33}{4}$

$$\sqrt{\frac{1089}{16} - \frac{264}{4} + 3} + 6 = \frac{33}{4}$$

$$\sqrt{\frac{81}{16}} + \frac{24}{4} = \frac{33}{4}$$

$$\frac{9}{4} + \frac{24}{4} = \frac{33}{4}$$

$$\frac{33}{4} = \frac{33}{4}$$

10.
(24)

$$\frac{5x - 2}{2} - \frac{x + 3}{3} = 2$$

$$15x - 6 - 2x - 6 = 12$$

$$13x = 24$$

$$x = \frac{24}{13}$$

11.
(56)

(a) $y = \dfrac{(360 - 100 - 90) - 100}{2} = \dfrac{70}{2} = \mathbf{35}$

(b) $y = \dfrac{110 + 90}{2} = \mathbf{100}$

12.
(B)

$$A_{\text{Base}} = 15(8) + 2\left(\frac{1}{2}\right)(\pi)(4)^2$$

$$= 120 + 16\pi \approx 170.24 \text{ ft}^2$$

$$V = A_{\text{Base}} \times H$$

$$\approx 170.24(8) \approx \mathbf{1361.92 \text{ ft}^3}$$

13.
(26)

$$\frac{3}{6} = \frac{h}{4}$$

$$6h = 4(3)$$

$$h = 2$$

$$x^2 = 2^2 + 3^2 = 4 + 9 = 13$$

$$x = \sqrt{13}$$

14.
(23)

(a) $6x - 4y = 14$

(a′) $y = \dfrac{3}{2}x - \dfrac{7}{2}$

(b) $3x + 2y = -1$

(b′) $y = -\dfrac{3}{2}x - \dfrac{1}{2}$

Substitute (a′) into (b′) to get:

(b′) $\dfrac{3}{2}x - \dfrac{7}{2} = -\dfrac{3}{2}x - \dfrac{1}{2}$

$$3x = 3$$

$$x = 1$$

(a′) $y = \dfrac{3}{2}(1) - \dfrac{7}{2} = -2$

(1, –2)

15.
(47)

$$\sqrt[5]{27\sqrt{3}} = \left[3^3(3)^{1/2}\right]^{1/5} = (3^{7/2})^{1/5} = \mathbf{3^{7/10}}$$

16.
(51)

$$4i^2 - \sqrt{-16} + 3\sqrt{-9} - 4 + \sqrt{49}$$

$$= 4(-1) - 4i + 3(3i) - 4 + 7$$

$$= -4 - 4i + 9i + 3 = \mathbf{-1 + 5i}$$

17.
(46)

$$\sqrt[4]{m^3 a^5}\,\sqrt[3]{m^5 a^4} = (m^3 a^5)^{1/4}(m^5 a^4)^{1/3}$$

$$= m^{3/4}a^{5/4}m^{5/3}a^{4/3} = \mathbf{m^{29/12}a^{31/12}}$$

18.
(35)
$$-16^{-5/4} = -\frac{1}{16^{5/4}} = -\frac{1}{(16^{1/4})^5}$$

$$= -\frac{1}{2^5} = -\frac{1}{32}$$

19.
(33)
$$\frac{\dfrac{bc}{m} - \dfrac{em}{b^2}}{\dfrac{4}{b^2} - \dfrac{bc}{b^2m}} = \frac{b^3c - em^2}{b^2m} \cdot \frac{b^2m}{4m - bc} \cdot \frac{\dfrac{b^2m}{4m - bc}}{\dfrac{b^2m}{4m - bc}}$$

$$= \frac{b^3c - em^2}{4m - bc}$$

20.
(36)
$$\frac{pqx^2 - 2pxq - 15qp}{-3x - 18 + x^2} \div \frac{-pqx + px^2q - 20qp}{-5x + x^2 - 6}$$

$$= \frac{\dfrac{pqx^2 - 2pxq - 15qp}{-3x - 18 + x^2}}{\dfrac{-5x + x^2 - 6}{-pqx + px^2q - 20qp}}$$

$$= \frac{pq(x - 5)(x + 3)}{(x - 6)(x + 3)} \cdot \frac{(x - 6)(x + 1)}{pq(x - 5)(x + 4)}$$

$$= \frac{x + 1}{x + 4}$$

TEST 16

1.
(57)
$$\frac{P_1}{T_1} = \frac{P_2}{T_2}$$

$$\frac{8000}{1500} = \frac{P_2}{3270}$$

$$1500P_2 = 8000(3270)$$

$$P_2 = \textbf{17,440 pascals}$$

2.
(61)
Alcohol$_1$ + Alcohol added = Alcohol final

$$A_1 + A_A = A_F$$

$$(360) + (P_N) = (360 + P_N)$$

$$0.3(360) + 1(P_N) = 0.4(360 + P_N)$$

$$108 + P_N = 144 + 0.4\,P_N$$

$$0.6\,P_N = 36$$

$$P_N = \textbf{60 mL}$$

3.
(60)
$$T = kP$$

$$15 = k(90)$$

$$\frac{1}{6} = k$$

$$T = \frac{1}{6}(102) = \textbf{17 test questions}$$

4.
(59)
$$W = m\mathrm{I}r + b$$

Use the graph to find the slope.

$$m = \frac{62.5 - 15}{0 - 10} = \frac{47.5}{-10} = -4.75$$

By inspection, $b = 62.5$.

$$\mathbf{W = -4.75Ir + 62.5}$$

5.
(59)
(a) $\dfrac{3}{8}x - \dfrac{1}{2}y = 1$

(b) $0.2x + 0.6y = 4$

(a′) $3x - 4y = 8$

(b′) $2x + 6y = 40$

$3($a′$)$ $9x - 12y = 24$

$2($b′$)$ $\underline{4x + 12y = 80}$

$13x = 104$

$x = 8$

(b′) $2(8) + 6y = 40$

$$6y = 24$$

$$y = 4$$

(8, 4)

6.
(54)
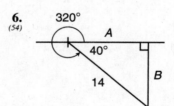

$$A = 14 \cos 40° \approx 10.72$$

$$B = 14 \sin 40° \approx 9.00$$

$$\mathbf{10.72R - 9.00U}$$

7.
(59)
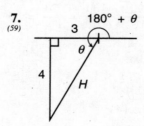

$$\tan \theta = \frac{4}{3}$$

$$\theta \approx 53.13$$

Add θ to 180° to get a third-quadrant angle:

$180° + 53.13° \approx 233.13°$

$H = \sqrt{3^2 + 4^2} = 5$

$5\underline{/233.13°}$

8.
(58)

$3x^2 + 5x - 4 = 0$

$\left(x^2 + \dfrac{5}{3}x + \quad\right) = \dfrac{4}{3}$

$x^2 + \dfrac{5}{3}x + \dfrac{25}{36} = \dfrac{48}{36} + \dfrac{25}{36}$

$\left(x + \dfrac{5}{6}\right)^2 = \dfrac{73}{36}$

$x + \dfrac{5}{6} = \pm\dfrac{\sqrt{73}}{6}$

$x = -\dfrac{5}{6} \pm \dfrac{\sqrt{73}}{6}$

9.
(38)

$56x + x^3 = 15x^2$

$x^3 - 15x^2 + 56x = 0$

$x(x^2 - 15x + 56) = 0$

$x(x - 7)(x - 8) = 0$

$x = \mathbf{0, 7, 8}$

10.
(24)

$\dfrac{3x - 4}{2} - \dfrac{3}{5} = 7$

$15x - 20 - 6 = 70$

$15x = 96$

$x = \dfrac{\mathbf{32}}{\mathbf{5}}$

11.
(55)

$\dfrac{4x}{2y + 3a} - p = \dfrac{c}{q}$

$4xq - 2pqy - 3apq = 2cy + 3ac$

$4xq - 2pqy - 2cy = 3ac + 3apq$

$\dfrac{\mathbf{4xq - 2pqy - 2cy}}{\mathbf{3c + 3pq}} = a$

12.
(53)

$300 \text{ in.}^3 \times \dfrac{2.54 \text{ cm}}{1 \text{ in.}} \times \dfrac{2.54 \text{ cm}}{1 \text{ in.}} \times \dfrac{2.54 \text{ cm}}{1 \text{ in.}}$

$\times \dfrac{1 \text{ m}}{100 \text{ cm}} \times \dfrac{1 \text{ m}}{100 \text{ cm}} \times \dfrac{1 \text{ m}}{100 \text{ cm}}$

$= \dfrac{\mathbf{300(2.54)(2.54)(2.54)}}{\mathbf{(100)(100)(100)}} \textbf{ m}^3$

13.
(1,17)

$\dfrac{3}{7} \times \overrightarrow{SF} = \dfrac{8}{21}$

$\overrightarrow{SF} = \dfrac{8}{9}$

$a \times \dfrac{8}{9} = \dfrac{1}{4}$

$a = \dfrac{\mathbf{9}}{\mathbf{32}}$

(a) $4x - 2y - 14 = 50$

(b) $5x - 4y + 38 = 130$

-2(a) $-8x + 4y + 28 = -100$

  (b) $\underline{5x - 4y + 38 = 130}$

$-3x + 66 = 30$

$-3x = -36$

$x = \mathbf{12}$

(a) $4(12) - 2y - 14 = 50$

$-2y = 16$

$y = \mathbf{-8}$

14.
(22)

$\dfrac{6}{8} = \dfrac{12}{Y}$

$6Y = 96$

$Y = 16$

$Z^2 = 16^2 + 8^2$

$Z^2 = 320$

$Z = \mathbf{8\sqrt{5}}$

15.
(31)

Write the equation of the given line in slope-intercept form.

$2x - 3y = 6$

$y = \dfrac{2}{3}x - 2$

Since the slopes of perpendicular lines are negative reciprocals of each other,

$m\perp = -\dfrac{3}{2}.$

$y = -\dfrac{3}{2}x + b$

$5 = -\dfrac{3}{2}(7) + b$

$\dfrac{31}{2} = b$

$y = -\dfrac{3}{2}x + \dfrac{31}{2}$

TEST 17

16. (a) $3x - y = -3$
(23)
(a') $y = 3x + 3$

(b) $x + y = 6$

$y = -x + 6$

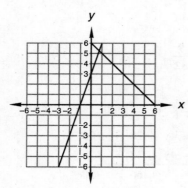

(a) $3x - y = -3$

(b) $\dfrac{x + y = 6}{4x \qquad = 3}$

$x = \dfrac{3}{4}$

(a') $y = 3\left(\dfrac{3}{4}\right) + 3 = \dfrac{21}{4}$

$\left(\dfrac{3}{4}, \dfrac{21}{4}\right)$

17. $-3i^2 + 4i - 5\sqrt{-9} - 4 + 5i^3$
(51)
$= -3(ii) + 4i - 5(3i) - 4 + 5i(ii)$

$= 3 + 4i - 15i - 4 - 5i = \mathbf{-1 - 16i}$

18. $3\sqrt{\dfrac{10}{11}} - 4\sqrt{\dfrac{11}{10}} - 5\sqrt{440}$
(46)

$= \dfrac{3\sqrt{10}}{\sqrt{11}} \cdot \dfrac{\sqrt{11}}{\sqrt{11}} - \dfrac{4\sqrt{11}}{\sqrt{10}} \cdot \dfrac{\sqrt{10}}{\sqrt{10}} - 5\sqrt{4}\sqrt{110}$

$= \dfrac{3\sqrt{110}}{11} - \dfrac{2\sqrt{110}}{5} - 10\sqrt{110}$

$= \dfrac{15\sqrt{110}}{55} - \dfrac{22\sqrt{110}}{55} - \dfrac{550\sqrt{110}}{55}$

$= \mathbf{-\dfrac{557\sqrt{110}}{55}}$

19. $\sqrt{20\sqrt{5}} = \sqrt{2^2 5\sqrt{5}} = [2^2(5)(5^{1/2})]^{1/2}$
(47)
$= 2(5^{1/2})(5^{1/4}) = \mathbf{2(5^{3/4})}$

20. $\dfrac{1}{(-32)^{-4/5}} = (-32)^{4/5} = ((-32)^{1/5})^4 = (-2)^4$
(35)
$= \mathbf{16}$

1. $\text{Al} = \dfrac{k}{\text{Sn}}$
(60)

$600 = \dfrac{k}{40}$

$24{,}000 = k$

$\text{Al} = \dfrac{24{,}000}{120} = \mathbf{200\ grams}$

2. $\dfrac{P_1 V_1}{T_1} = \dfrac{P_2 V_2}{T_2}$
(57)

$\dfrac{(450)(4)}{300} = \dfrac{P_2(12)}{500}$

$P_2 = \mathbf{250\ N/m^2}$

3. Protein P_N + Protein D_N = Protein Total
(52)
$0.24 P_N + 0.12(D_N) = 0.16(900)$

(a) $0.24 P_N + 0.12 D_N = 144$

(b) $P_N + D_N = 900$

(b') $D_N = 900 - P_N$

Substitute (b') into (a') to get:

(a') $0.24 P_N + 0.12(900 - P_N) = 144$

$0.12 P_N + 108 = 144$

$0.12 P_N = 36$

$P_N = \mathbf{300\ pounds\ of\ 24\%\ solution}$

(b') $D_N = 900 - 300$

$= \mathbf{600\ pounds\ of\ 12\%\ solution}$

4.
(59)

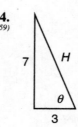

$\tan \theta = \dfrac{7}{3}$

$\theta \approx 66.80°$

Subtract θ from 180° to get a second-quadrant angle:

$\theta \approx 180° - 66.80° \approx 113.20°$

$H = \sqrt{3^2 + 7^2} = \sqrt{58}$

$\mathbf{\sqrt{58}\ \underline{/113.20°}}$

5.
(63)

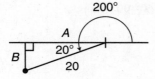

$A = 20 \cos 20° \approx 18.79$

$B = 20 \sin 20° \approx 6.84$

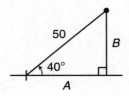

$A = 50 \cos 40° \approx 38.30$

$B = 50 \sin 40° \approx 32.14$

$$-18.79R - 6.84U$$
$$\underline{38.30R + 32.14U}$$
$$\mathbf{19.51R + 25.30U}$$

6.
(59)

(a) $\dfrac{2}{3}x + \dfrac{2}{5}y = 18$

(b) $-0.1x - 0.2y = -5.5$

   $-x - 2y = -55$

(b′) $x = -2y + 55$

Substitute (b′) into (a) to get:

(a′) $\dfrac{2}{3}(-2y + 55) + \dfrac{2}{5}y = 18$

$-20y + 550 + 6y = 270$

$-14y = -280$

$y = 20$

(b′) $x = -2(20) + 55 = 15$

(15, 20)

7.
(62)

$4 = 2x - 3x^2$

$-\dfrac{4}{3} = \left(x^2 - \dfrac{2}{3}x + \phantom{\dfrac{1}{9}} \right)$

$\dfrac{1}{9} - \dfrac{4}{3} = x^2 - \dfrac{2}{3}x + \dfrac{1}{9}$

$-\dfrac{11}{9} = \left(x - \dfrac{1}{3} \right)^2$

$\pm\dfrac{\sqrt{11}}{3}i = x - \dfrac{1}{3}$

$x = \dfrac{1}{3} \pm \dfrac{\sqrt{11}}{3}i$

8.
(4)

$-9^0 - 3^2 - 2^3(-3^0 - 3)x - 4x - 13x^0$

$= -x^0 + 3$

$-1 - 9 - 8(-4)x - 4x - 13 = -1 + 3$

$28x - 23 = 2$

$28x = 25$

$x = \dfrac{25}{28}$

9.
(B)

$H^2 = 5^2 + 12^2$

$H^2 = 169$

$H = 13$

$S.A. = \dfrac{2(10 \times 12)}{2} + 2(25 \times 13) + (25 \times 10)$

$= 120 + 650 + 250 = 1020 \text{ ft}^2$

$1020 \text{ ft}^2 \times \dfrac{12 \text{ in.}}{1 \text{ ft}} \times \dfrac{12 \text{ in.}}{1 \text{ ft}} = \mathbf{146{,}880 \text{ in.}^2}$

10.
(59)

$H = mC + b$

Use the graph to find the slope.

Slope $= \dfrac{130}{20} = 6.5$

$H = 6.5C + b$

Use (20, 140) for C and H.

$140 = 6.5(20) + b$

$b = 10$

$H = \mathbf{6.5C + 10}$

11.
(55)

$\dfrac{a}{b + c} - d = \dfrac{e}{f}$

$af - dbf - dcf = eb + ec$

$af - dbf - eb = dcf + ec$

$af - dbf - eb = c(df + e)$

$\dfrac{af - dbf - eb}{df + e} = c$

12.
(64)

$\sqrt{-6}\sqrt{-7} - \sqrt{-25} - \sqrt{-3}\sqrt{-3} + 5i^7$

$= \sqrt{6}i\sqrt{7}i - 5i - \sqrt{3}i\sqrt{3}i + 5i(ii)(ii)(ii)$

$= \mathbf{3 - \sqrt{42} - 10i}$

13. $(3i + 7)(2 - 5i) = 6i + 14 - 15i^2 - 35i$
(64)

$= \textbf{29} - \textbf{29}\textbf{\textit{i}}$

14. $\sqrt[4]{x^{1/3}\sqrt{x^5}} = [x^{1/3}(x^{5/2})]^{1/4} = (x^{17/6})^{1/4}$
(46)

$= \textbf{\textit{x}}^{\textbf{17/24}}$

15. $\sqrt{8\sqrt{2}} = [2^3(2)^{1/2}]^{1/2} = (2^{7/2})^{1/2} = \textbf{2}^{\textbf{7/4}}$
(47)

16. $\dfrac{2x + 3}{x - 5} - \dfrac{5x - 4}{5 - x} = \dfrac{2x + 3}{x - 5} + \dfrac{5x - 4}{x - 5}$
(66)

$= \dfrac{\textbf{7}\textbf{\textit{x}} - \textbf{1}}{\textbf{\textit{x}} - \textbf{5}}$

17. (a) $R_B T_B = 480$
(65)

(b) $R_F T_F = 160$

(c) $R_B = 6R_F$

(d) $T_B + T_F = 6$

(d′) $T_B = 6 - T_F$

Substitute (d′) and (c) into (a) to get:

(a′) $6R_F(6 - T_F) = 480$

Substitute (b) into (a′) to get:

(a″) $36R_F - 6(160) = 480$

$R_F = \textbf{40}$

(c) $R_B = 6(40) = \textbf{240}$

(b) $40T_F = 160$

$T_F = \textbf{4}$

(d′) $T_B = 6 - 4 = \textbf{2}$

18. $\dfrac{90 \text{ in.}^3}{\text{min}} \times \dfrac{2.54 \text{ cm}}{1 \text{ in.}} \times \dfrac{2.54 \text{ cm}}{1 \text{ in.}} \times \dfrac{2.54 \text{ cm}}{1 \text{ in.}}$
(53)

$\times \dfrac{1 \text{ m}}{100 \text{ cm}} \times \dfrac{1 \text{ m}}{100 \text{ cm}} \times \dfrac{1 \text{ m}}{100 \text{ cm}} \times \dfrac{1 \text{ min}}{60 \text{ s}}$

$= \dfrac{\textbf{90(2.54)(2.54)(2.54)}}{\textbf{(100)(100)(100)(60)}} \dfrac{\textbf{m}^{\textbf{3}}}{\textbf{s}}$

19. $\dfrac{a}{m} + \dfrac{5}{3 + \dfrac{2a}{m}} = \dfrac{a}{m} + \dfrac{5}{\dfrac{3m + 2a}{m}}$
(64)

$= \dfrac{a}{m} + \dfrac{5m}{3m + 2a}$

$= \dfrac{a(3m + 2a)}{m(3m + 2a)} + \dfrac{5m^2}{m(3m + 2a)}$

$= \dfrac{\textbf{5}\textbf{\textit{m}}^{\textbf{2}} + \textbf{3}\textbf{\textit{ma}} + \textbf{2}\textbf{\textit{a}}^{\textbf{2}}}{\textbf{\textit{m}}(\textbf{3}\textbf{\textit{m}} + \textbf{2}\textbf{\textit{a}})}$

20. Since this is a 30°-60°-90° triangle:
(66)

$2 \times \overrightarrow{SF} = 7$

$\overrightarrow{SF} = \dfrac{7}{2}$

$a = \dfrac{7}{2}(\sqrt{3}) = \dfrac{\textbf{7}\sqrt{\textbf{3}}}{\textbf{2}}$

$b = \dfrac{7}{2}(1) = \dfrac{\textbf{7}}{\textbf{2}}$

TEST 18

1. $\dfrac{P_1 V_1}{T_1} = \dfrac{P_2 V_2}{T_2}$
(69)

$T_2 = \dfrac{P_2 V_2 T_1}{P_1 V_1}$

$T_2 = \dfrac{(3000 \times 10^3)(30{,}000)(7000)}{(0.03 \times 10^9)(0.007 \times 10^{-2})}$

$= \dfrac{6.3 \times 10^{14}}{2.1 \times 10^3} = \textbf{3} \times \textbf{10}^{\textbf{11}} \textbf{ K}$

2. Saline P_N + Saline D_N = Saline Total
(52)

$0.90(P_N) + 0.40(D_N) = (0.60)(2000)$

(a) $0.9P_N + 0.4D_N = 1200$

(b) $P_N + D_N = 2000$

(b′) $D_N = 2000 - P_N$

Substitute (b′) into (a) to get:

(a′) $0.9P_N + 0.4(2000 - P_N) = 1200$

$0.5P_N + 800 = 1200$

$0.5P_N = 400$

$P_N = \textbf{800 L of}$
$\textbf{90\% solution}$

(b′) $D_N = 2000 - 800$

$= \textbf{1200 L of 40\% solution}$

3. $N \quad N + 1 \quad N + 2$
(55)

$3(N)(N + 2) = 2(N)(N + 1) + 32$

$3N^2 + 6N = 2N^2 + 2N + 32$

$N^2 + 4N - 32 = 0$

$(N + 8)(N - 4) = 0$

$N = -8, 4$

Since we are looking for consecutive negative integers, the desired integers are **−8, −7,** and **−6.**

4.
(67)
$$\frac{3}{3\sqrt{2} + 5} \cdot \frac{3\sqrt{2} - 5}{3\sqrt{2} - 5} = \frac{9\sqrt{2} - 15}{18 - 25}$$

$$= \frac{15}{7} - \frac{9\sqrt{2}}{7}$$

5.
(46)
$$6\sqrt{\frac{7}{5}} - 2\sqrt{\frac{5}{7}} - 3\sqrt{140}$$

$$= \frac{6\sqrt{7}}{\sqrt{5}} \cdot \frac{\sqrt{5}}{\sqrt{5}} - \frac{2\sqrt{5}}{\sqrt{7}} \cdot \frac{\sqrt{7}}{\sqrt{7}} - 6\sqrt{35}$$

$$= \frac{6\sqrt{35}}{5} - \frac{2\sqrt{35}}{7} - 6\sqrt{35}$$

$$= \frac{42\sqrt{35}}{35} - \frac{10\sqrt{35}}{35} - \frac{210\sqrt{35}}{35} = -\frac{178\sqrt{35}}{35}$$

6.
(46)
$$\sqrt[4]{x^2 c^5}\,\sqrt[5]{xc^4} = (x^2 c^5)^{1/4}(xc^4)^{1/5}$$
$$= x^{1/2}c^{5/4}x^{1/5}c^{4/5} = x^{7/10}c^{41/20}$$

7.
(27)
$$\frac{3}{x + 3} - \frac{2x + 1}{x^2 - 9}$$

$$= \frac{3x - 9}{(x + 3)(x - 3)} - \frac{2x + 1}{(x + 3)(x - 3)}$$

$$= \frac{x - 10}{(x + 3)(x - 3)}$$

8.
(63)

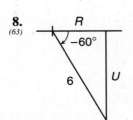

$$R = 6 \cos 60° = 3$$
$$U = 6 \sin 60° \approx 5.20$$

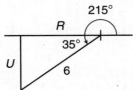

$$R = 6 \cos 35° \approx 4.91$$
$$U = 6 \sin 35° \approx 3.44$$

$$3.00R - 5.20U$$
$$-4.91R - 3.44U$$
$$\overline{-1.91R - 8.64U}$$

9.
(59)

$$\tan \theta = \frac{2}{3}$$

$$\theta \approx 33.69°$$

Subtract θ from 360° to get a fourth-quadrant angle:

$$360° - 33.69° \approx 326.31°$$

$$H = \sqrt{3^2 + (-2)^2} = \sqrt{13}$$

$$\sqrt{13}\,\underline{/326.31°}$$

10.
(65)
(a) $R_M T_M = 240$

(b) $R_Q T_Q = 240$

(c) $R_M = 2R_Q$

(d) $T_M = T_Q - 4$

Substitute (c) and (d) into (a) to get:

$$2R_Q(T_Q - 4) = 240$$

(a′) $2R_Q T_Q - 8R_Q = 240$

Substitute (b) into (a′) to get:

(a″) $2(240) - 8R_Q = 240$

$$-8R_Q = -240$$

$$R_Q = 30$$

(c) $R_M = 2(30) = 60$

(b) $30T_Q = 240$

$$T_Q = 8$$

(d) $T_M = 8 - 4 = 4$

11.
(64)
$$x + \frac{3b}{x + \frac{2}{b}} = x + \frac{3b}{\frac{bx + 2}{b}} = x + \frac{3b^2}{bx + 2}$$

$$= \frac{bx^2 + 2x + 3b^2}{bx + 2}$$

12.
(64)
$$(2i - 5)(4i - 3) - \sqrt{-5}\,\sqrt{-5} + 6i^2 - i^3$$
$$= -8 - 20i - 6i + 15 + 5 - 6 + i = 6 - 25i$$

13. (a) $\dfrac{1}{5}x - \dfrac{5}{2}y = -48$
(59)

$2x - 25y = -480$

(a') $x = \dfrac{25}{2}y - 240$

(b) $0.4x + 0.05y = 5$

(b') $40x + 5y = 500$

Substitute (a') into (b') to get:

(b'') $40\left(\dfrac{25}{2}y - 240\right) + 5y = 500$

$505y - 9600 = 500$

$505y = 10{,}100$

$y = 20$

(a') $x = \dfrac{25}{2}(20) - 240 = 10$

(10, 20)

14. $x + 6 = 2x^2$
(58)

$3 = \left(x^2 - \dfrac{1}{2}x + \quad\right)$

$3 + \dfrac{1}{16} = x^2 - \dfrac{1}{2}x + \dfrac{1}{16}$

$\dfrac{49}{16} = \left(x - \dfrac{1}{4}\right)^2$

$\pm\dfrac{7}{4} = x - \dfrac{1}{4}$

$x = \dfrac{1}{4} \pm \dfrac{7}{4} = \mathbf{2, -\dfrac{3}{2}}$

15. $\dfrac{4x + 5}{3} + 2 = \dfrac{x}{7}$
(24)

$28x + 35 + 42 = 3x$

$77 = -25x$

$-\dfrac{77}{25} = x$

16. $\dfrac{300 \text{ ft}^3}{\text{min}} \times \dfrac{1 \text{ yd}}{3 \text{ ft}} \times \dfrac{1 \text{ yd}}{3 \text{ ft}} \times \dfrac{1 \text{ yd}}{3 \text{ ft}} \times \dfrac{1 \text{ min}}{60 \text{ s}}$
(47)

$= \dfrac{300}{(3)(3)(3)(60)} \dfrac{\text{yd}^3}{\text{s}}$

17. $ay = b\left(\dfrac{c}{x + e} + \dfrac{3f}{h}\right)$
(70)

$ay = \dfrac{bc}{x + e} + \dfrac{3bf}{h}$

$xhay + ehay = bch + 3bfx + 3bfe$

$xhay - 3bfx = bch + 3bfe - ehay$

$x = \dfrac{bch + 3bfe - ehay}{hay - 3bf}$

18. Write the equation of the given line in slope-
(20) intercept form.

$4x - 3y = 12$

$y = \dfrac{4}{3}x - 4$

$y = \dfrac{4}{3}x + b$

$-5 = \dfrac{4}{3}(3) + b$

$-9 = b$

$y = \dfrac{4}{3}x - 9$

19. (a) $\dfrac{373{,}402 \times 10^{10}}{97{,}376 \times 10^{-4}} \approx \dfrac{4 \times 10^{15}}{1 \times 10^{1}} \approx 4 \times 10^{14}$
(68)

Calculator $\approx \mathbf{3.83 \times 10^{14}}$

(b) $\sqrt[4.3]{207} \longrightarrow 3^4 = 81$ and $4^4 = 256$

Answer should be between 3 and 4.

Calculator $\approx \mathbf{3.46}$

20. Since $\triangle ABC$ is isosceles
(11,13)

$m\widehat{AC} = m\widehat{BC}$

$100 + 2m\widehat{AC} = 360$

$2m\widehat{AC} = 260$

$m\widehat{AC} = \mathbf{130°}$

$m\widehat{BC} = \mathbf{130°}$

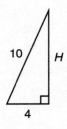

$10^2 = 4^2 + H^2$

$84 = H^2$

$2\sqrt{21} = H$

$A = \dfrac{1}{2}(8)(2\sqrt{21}) = \mathbf{8\sqrt{21}} \text{ units}^2$

TEST 19

1.
(57)

$$\frac{P_1}{T_1} = \frac{P_2}{T_2}$$

$$\frac{1600}{240} = \frac{1200}{T_2}$$

$$T_2 = \textbf{180 K}$$

2.
(74)

$R_A T_A = 60; \quad R_C T_C = 32$

$R_A = 3R_C; \quad T_A = T_C - 3$

$3R_C(T_C - 3) = 60$

$3R_C T_C - 9R_C = 60$

$96 - 9R_C = 60$

$-9R_C = -36$

$R_C = \textbf{4 mph}$

$R_A = 3(4) = \textbf{12 mph}$

$(4)T_C = 32$

$T_C = \textbf{8 hours}$

$T_A = 8 - 3 = \textbf{5 hours}$

3.
(21)

(a) $3N_B = 2N_R + 18$

(b) $5N_B = 6N_R - 2$

$3(a) \quad 9N_B = 6N_R + 54$

$-(b) \quad \underline{-5N_B = -6N_R + 2}$

$\qquad 4N_B = + 56$

$\qquad N_B = \textbf{14 blue marbles}$

(a) $3(14) = 2N_R + 18$

$24 = 2N_R$

$N_R = \textbf{12 red marbles}$

4.
(71)

$$ax^2 + bx + c = 0$$

$$\left(x^2 + \frac{b}{a}x + \phantom{\frac{b^2}{4a^2}}\right) = -\frac{c}{a}$$

$$x^2 + \frac{b}{a}x + \frac{b^2}{4a^2} = -\frac{c}{a} + \frac{b^2}{4a^2}$$

$$\left(x + \frac{b}{2a}\right)^2 = \frac{b^2 - 4ac}{4a^2}$$

$$x + \frac{b}{2a} = \pm\frac{\sqrt{b^2 - 4ac}}{2a}$$

$$x = \frac{-b \pm \sqrt{b^2 - 4ac}}{2a}$$

5.
(71)

$$2x^2 + 3 = -2x$$

$$2x^2 + 2x + 3 = 0$$

$$x = \frac{-2 \pm \sqrt{(2)^2 - 4(2)(3)}}{2(2)}$$

$$= \frac{-2 \pm \sqrt{-20}}{2(2)}$$

$$= -\frac{1}{2} \pm \frac{\sqrt{5}i}{2}$$

6.
(56)

(a) $x = \dfrac{180 - (360 - 180 - 130)}{2}$

$= \dfrac{180 - 50}{2} = \textbf{65}$

(b) $x = \dfrac{110}{2} = \textbf{55}$

7.
(70)

$$xk = p\left(\frac{1}{cm_1} - \frac{1}{m_2}\right)$$

$$xk = \frac{p}{cm_1} - \frac{p}{m_2}$$

$$cm_1 m_2 xk = pm_2 - cpm_1$$

$$cpm_1 + cm_1 m_2 xk = pm_2$$

$$c(pm_1 + m_1 m_2 xk) = pm_2$$

$$c = \frac{\textbf{\textit{pm}}_2}{\textbf{\textit{pm}}_1 + \textbf{\textit{m}}_1\textbf{\textit{m}}_2\textbf{\textit{xk}}}$$

8.
(27)

$$\frac{2x + 1}{x + 4} + \frac{3x - 2}{(x + 4)(x - 1)}$$

$$= \frac{(2x + 1)(x - 1)}{(x + 4)(x - 1)} + \frac{3x - 2}{(x + 4)(x - 1)}$$

$$= \frac{2x^2 + x - 2x - 1 + 3x - 2}{(x + 4)(x - 1)}$$

$$= \frac{\textbf{2}\textit{x}^2 + \textbf{2}\textit{x} - \textbf{3}}{(\textit{x} + \textbf{4})(\textit{x} - \textbf{1})}$$

9.
(73)

$$\frac{3\sqrt{2} - 4}{\sqrt{3} - 2} \cdot \frac{\sqrt{3} + 2}{\sqrt{3} + 2}$$

$$= \frac{3\sqrt{6} - 4\sqrt{3} + 6\sqrt{2} - 8}{3 - 4}$$

$$= \textbf{8} - \textbf{6}\sqrt{\textbf{2}} + \textbf{4}\sqrt{\textbf{3}} - \textbf{3}\sqrt{\textbf{6}}$$

10.
(47)

$-2\sqrt{4\sqrt{2}} = -2[2^2(2)^{1/2}]^{1/2} = (-2)(2)(2^{1/4})$

$= -\textbf{2}^{\textbf{9/4}}$

11.
(35)

$\dfrac{-5^2}{-27^{-2/3}} = 25(27^{1/3})^2 = 25(3)^2 = \textbf{225}$

12.
(64)
$$ax + \cfrac{a}{x - \cfrac{x}{a}} = ax + \cfrac{a}{\cfrac{ax - x}{a}}$$

$$= ax + \cfrac{a^2}{ax - x}$$

$$= \cfrac{a^2x^2 - ax^2 + a^2}{ax - x}$$

13.
(64)
$$(3i + 2)(2i - 3) - \sqrt{-16} - 2i^2 + i^3$$

$$= 6i^2 + 4i - 9i - 6 - 4i - 2(-1) + i(ii)$$

$$= -10 - 10i$$

14.
(46)
$$\sqrt{a^3x^0y^2a^{1/2}x^3y} = (a^{7/2}x^3y^3)^{1/2} = a^{7/4}x^{3/2}y^{3/2}$$

15. Graph the line to find the slope.
(31)

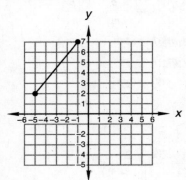

$$m = \frac{5}{4}$$

Since the slopes of perpendicular lines are negative reciprocals of each other,

$$m\perp = -\frac{4}{5}.$$

$$y = -\frac{4}{5}x + b$$

$$-1 = -\frac{4}{5}(3) + b$$

$$\frac{7}{5} = b$$

$$y = -\frac{4}{5}x + \frac{7}{5}$$

16.
(72)

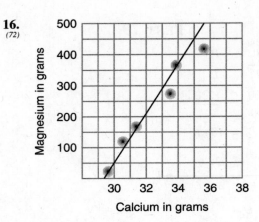

Use the graph to find the slope.

$$m = \frac{400}{5} = 80$$

Mg = 80Ca + b

Use (30, 50) for Calcium and Magnesium.

$$50 = 80(30) + b$$

$$-2350 = b$$

Mg = 80Ca − 2350

17. (a)
(59,72)

$$\tan \theta = \frac{3}{3}$$

$$\theta = 45°$$

Subtract θ from 180° to get a second-quadrant angle:

$$\theta = 180° - 45° = 135°$$

$$H = \sqrt{3^2 + 3^2} = 3\sqrt{2}$$

$$3\sqrt{2}\,\underline{/135°}$$

(b)

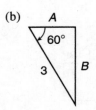

$$A = 3 \cos 60° = 1.50$$

$$B = 3 \sin 60° \approx 2.60$$

1.50R − 2.60U

18.
(62)

$$2x^2 + 3 = x$$

$$\left(x^2 - \frac{1}{2}x + \quad\right) = -\frac{3}{2}$$

$$x^2 - \frac{1}{2}x + \frac{1}{16} = -\frac{3}{2} + \frac{1}{16}$$

$$\left(x - \frac{1}{4}\right)^2 = -\frac{23}{16}$$

$$x - \frac{1}{4} = \pm\frac{\sqrt{23}}{4}i$$

$$x = \mathbf{\frac{1}{4} \pm \frac{\sqrt{23}}{4}i}$$

19.
(48)

$$3 + \sqrt{4x - 5} = 12$$

$$\sqrt{4x - 5} = 9$$

$$4x - 5 = 81$$

$$4x = 86$$

$$x = \mathbf{\frac{43}{2}}$$

Check: $3 + \sqrt{4\left(\dfrac{43}{2}\right) - 5} = 12$

$$3 + \sqrt{81} = 12$$

$$12 = 12$$

20.
(26)

$$X^2 = 7^2 + 5^2 = 49 + 25 = 74$$

$$X = \sqrt{74}$$

$$\frac{7}{12} = \frac{\sqrt{74}}{Y + \sqrt{74}}$$

$$7Y + 7\sqrt{74} = 12\sqrt{74}$$

$$Y = \mathbf{\frac{5\sqrt{74}}{7}}$$

TEST 20

1.
(74)

$$R_O T_O = R_B T_B = 350$$

$$R_O = 2R_B; \quad T_O = T_B - 5$$

$$2R_B(T_B - 5) = 350$$

$$2R_B T_B - 10R_B = 350$$

$$2(350) - 10R_B = 350$$

$$-10R_B = -350$$

$$R_B = \mathbf{35\ mph}$$

$$R_O = 2(35) = \mathbf{70\ mph}$$

$$35T_B = 350$$

$$T_B = \mathbf{10\ hours}$$

$$T_O = 10 - 5 = \mathbf{5\ hours}$$

2.
(52)

Protein C + Protein F = Protein Total

$$0.20(C) + 0.60(F) = 0.30(3600)$$

(a) $0.2C + 0.6F = 1080$

(b) $C + F = 3600$

(b′) $F = 3600 - C$

Substitute (b′) into (a) to get:

$$0.2C + 0.6(3600 - C) = 1080$$

$$2160 - 0.4C = 1080$$

$$0.4C = 1080$$

$$C = \mathbf{2700\ pounds}$$
$$\mathbf{of\ cornmeal}$$

(b′) $F = 3600 - 2700$

$$= \mathbf{900\ pounds\ of\ feed\ supplement}$$

3.
(21)

(a) $7N_L = 4N_S + 8$

(a′) $N_L = \dfrac{4}{7}N_S + \dfrac{8}{7}$

(b) $3N_S = 2N_L - 19$

Substitute (a′) into (b) to get:

(b′) $\quad 3N_S = 2\left(\dfrac{4}{7}N_S + \dfrac{8}{7}\right) - 19$

$$\frac{21}{7}N_S = \frac{8}{7}N_S + \frac{16}{7} - \frac{133}{7}$$

$$\frac{13}{7}N_S = -\frac{117}{7}$$

$$N_S = \mathbf{-9}$$

(a′) $N_L = \dfrac{4}{7}(-9) + \dfrac{8}{7} = -\dfrac{36}{7} + \dfrac{8}{7} = \mathbf{-4}$

4.
(59)

(a) $\dfrac{1}{5}x - \dfrac{1}{3}y = 1$

(a′) $3x - 5y = 15$

(b) $0.03x - 0.7y = -3.75$

(b′) $3x - 70y = -375$

$$\begin{array}{r} \text{(a′)} \quad 3x - 5y = 15 \\ -1\text{(b′)} \quad \underline{-3x + 70y = 375} \\ 65y = 390 \\ y = 6 \end{array}$$

(a′) $3x - 5(6) = 15$

$$3x = 45$$

$$x = 15$$

$(\mathbf{15, 6})$

5. (a) $x + y + z = 11$
(76)
(b) $2x + y - z = 13$

(c) $x = 2y$

Substitute (c) into (a) to get:

(a') $2y + y + z = 11$

$\qquad 3y + z = 11$

(a'') $z = 11 - 3y$

Substitute (c) and (a'') into (b) to get:

(b') $2(2y) + y - (11 - 3y) = 13$

$\qquad 8y - 11 = 13$

$\qquad 8y = 24$

$\qquad y = 3$

(c) $x = 2(3) = 6$

(a'') $z = 11 - 3(3) = 2$

(6, 3, 2)

6. $\sqrt{x - 25} + \sqrt{x} = 5$
(77)
$\qquad \sqrt{x - 25} = 5 - \sqrt{x}$

$\qquad x - 25 = 25 - 10\sqrt{x} + x$

$\qquad -50 = -10\sqrt{x}$

$\qquad 5 = \sqrt{x}$

$\qquad \mathbf{25 = x}$

Check: $\sqrt{25 - 25} + \sqrt{25} = 5$

$\qquad\qquad\qquad 5 = 5$

7. $\dfrac{4x - 3}{7} = 5 + \dfrac{x - 2}{3}$
(24)
$12x - 9 = 105 + 7x - 14$

$\qquad 5x = 100$

$\qquad x = \mathbf{20}$

8.
(78)

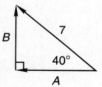

$A = 7 \cos 40° \approx 5.36$

$B = 7 \sin 40° \approx 4.50$

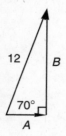

$A = 12 \cos 70° \approx 4.10$

$B = 12 \sin 70° \approx 11.28$

$\qquad -5.36R + \quad 4.50U$

$\qquad \underline{4.10R + 11.28U}$

$\qquad \mathbf{-1.26R + 15.78U}$

$\tan \theta \approx \dfrac{15.78}{1.26}$

$\qquad \theta \approx 85.43°$

Subtract θ from $180°$ to get a second-quadrant angle:

$180° - 85.43° \approx 94.57°$

$F \approx \sqrt{(-1.26)^2 + 15.78^2} \approx 15.83$

15.83$\underline{/94.57°}$

9. (a) $\dfrac{41{,}805 \times 10^3}{0.000395 \times 10^{-1}} \approx \dfrac{4 \times 10^7}{4 \times 10^{-5}} = 1 \times 10^{12}$
(68)
Calculator $\approx \mathbf{1.06 \times 10^{12}}$

(b) $\sqrt[2.3]{362} \longrightarrow 18^2 = 324$ and $19^2 = 361$

Answer might be between 18 and 19.

Calculator $\approx \mathbf{12.96}$

10.
(22)

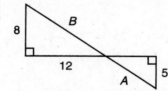

$B = \sqrt{8^2 + 12^2} = \sqrt{208} = 4\sqrt{13}$

$\dfrac{5}{8} = \dfrac{A}{4\sqrt{13}}$

$A = \dfrac{20\sqrt{13}}{8} = \dfrac{\mathbf{5\sqrt{13}}}{\mathbf{2}}$

11. $(3\sqrt{3})^2 = 3^2 + H^2$
(13)
$\qquad 27 = 9 + H^2$

$\qquad 18 = H^2$

$\qquad 3\sqrt{2} = H$

$A = \dfrac{1}{2}(6)(3\sqrt{2}) + \dfrac{1}{2}(\pi)\left(\dfrac{3\sqrt{3}}{2}\right)^2 \approx \mathbf{23.33\ cm^2}$

12. $\qquad 14 - 13x = -3x^2$
(71)
$3x^2 - 13x + 14 = 0$

$x = \dfrac{13 \pm \sqrt{(-13)^2 - 4(3)(14)}}{2(3)}$

$\dfrac{13 \pm \sqrt{1}}{6} = \mathbf{2, \dfrac{7}{3}}$

13.
(70)
$$\frac{a}{m} = c\left(\frac{1}{x} + \frac{b}{y}\right)$$

$$\frac{a}{m} = \frac{c}{x} + \frac{cb}{y}$$

$$axy = bcmx + cmy$$

$$axy - cmy = bcmx$$

$$\frac{axy - cmy}{cmx} = b$$

14.
(75)
$$\frac{2x + 7}{x^2 - x - 6} - \frac{4x}{3 - x}$$

$$= \frac{2x + 7}{(x - 3)(x + 2)} + \frac{4x^2 + 8x}{(x - 3)(x + 2)}$$

$$= \frac{4x^2 + 10x + 7}{(x - 3)(x + 2)}$$

15.
(73)
$$\frac{\sqrt{5} - 3}{\sqrt{5} - 2} \cdot \frac{\sqrt{5} + 2}{\sqrt{5} + 2} = \frac{5 - 3\sqrt{5} + 2\sqrt{5} - 6}{5 - 4}$$

$$= -1 - \sqrt{5}$$

16.
(46)
$$2\sqrt{\frac{5}{7}} + 3\sqrt{\frac{7}{5}} - 2\sqrt{140}$$

$$= \frac{2\sqrt{5}}{\sqrt{7}} \cdot \frac{\sqrt{7}}{\sqrt{7}} + \frac{3\sqrt{7}}{\sqrt{5}} \cdot \frac{\sqrt{5}}{\sqrt{5}} - 4\sqrt{35}$$

$$= \frac{2\sqrt{35}}{7} + \frac{3\sqrt{35}}{5} - 4\sqrt{35}$$

$$= \frac{10\sqrt{35}}{35} + \frac{21\sqrt{35}}{35} - \frac{140\sqrt{35}}{35} = -\frac{109\sqrt{35}}{35}$$

17.
(20)
$$(5\sqrt{16} - 3)(2\sqrt{10} + 2)$$
$$= 10\sqrt{2}\sqrt{2}\sqrt{2}\sqrt{2}\sqrt{2}\sqrt{5}$$
$$\quad - 6\sqrt{2}\sqrt{5} + 10(4) - 6$$
$$= 40\sqrt{10} - 6\sqrt{10} + 34 = 34 + 34\sqrt{10}$$

18.
(64)
$$ax^2 - \frac{a}{a + \frac{1}{ax}} = ax^2 - \frac{a}{\frac{a^2x + 1}{ax}}$$

$$= ax^2 - \frac{a^2x}{a^2x + 1} = \frac{a^3x^3 + ax^2 - a^2x}{a^2x + 1}$$

19.
(64)
$$\sqrt{-16} - \sqrt{-3}\sqrt{-3} - \sqrt{-4}\sqrt{-4} + 3i - 3i^2 + 3i^3$$
$$= 4i + 3 + 4 + 3i + 3 + 3i(i i) = 10 + 4i$$

20.
(35)
$$\frac{-2^0(-4^2)}{-4^{3/2}} = -(-2^4)\left[-(2^2)^{-3/2}\right] = -2^4(2^{-3})$$

$$= -2$$

TEST 21

1.
(80)
$$W_A = \frac{k}{(W_c)^2}$$

$$20 = \frac{k}{10^2}$$

$$2000 = k$$

$$W_A = \frac{2000}{5^2}$$

$$W_A = \textbf{80 kg}$$

2.
(74)

$$\begin{array}{c} D_P \\ \vdash\!\!\!\!\dashv\!\vdash\!\!\!\longrightarrow \\ 2080 \end{array}$$

$$\begin{array}{c} D_C \\ \vdash\!\!\!\dashv \\ 195 \end{array}$$

$$R_P T_P = 2080; \quad R_C T_C = 195$$
$$R_P = 4R_C; \quad T_P = T_C + 5$$
$$4R_C(T_C + 5) = 2080$$
$$4R_C T_C + 20R_C = 2080$$
$$780 + 20R_C = 2080$$
$$R_C = \textbf{65 mph}$$

$$65(T_C) = 195$$
$$T_C = \textbf{3 hours}$$
$$R_P = 4(65) = \textbf{260 mph}$$
$$T_P = 3 + 5 = \textbf{8 hours}$$

3.
(21)
(a) $\dfrac{8}{5} = \dfrac{N}{D} \longrightarrow 5N = 8D$

(a') $N = \dfrac{8}{5}D$

(b) $3N = 5D - 3$

Substitute (a') into (b) to get:

(b') $3\left(\dfrac{8}{5}D\right) = 5D - 3$

$$24D = 25D - 15$$
$$D = 15$$

(a') $N = \dfrac{8}{5}(15) = \textbf{24}$

4.
(53)
$$\frac{200 \text{ cm}^3}{s} \times \frac{1 \text{ in.}}{2.54 \text{ cm}} \times \frac{1 \text{ in.}}{2.54 \text{ cm}} \times \frac{1 \text{ in.}}{2.54 \text{ cm}}$$

$$\times \frac{1 \text{ ft}}{12 \text{ in.}} \times \frac{1 \text{ ft}}{12 \text{ in.}} \times \frac{1 \text{ ft}}{12 \text{ in.}} \times \frac{60}{1} \frac{s}{\min}$$

$$= \frac{200(60)}{(2.54)(2.54)(2.54)(12)(12)(12)} \frac{\text{ft}^3}{\min}$$

5.
(77)
$$\sqrt{x - 95} + \sqrt{x} = 19$$
$$\sqrt{x - 95} = 19 - \sqrt{x}$$
$$x - 95 = 361 - 38\sqrt{x} + x$$
$$-456 = -38\sqrt{x}$$
$$12 = \sqrt{x}$$
$$\mathbf{144 = x}$$
Check: $\sqrt{144 - 95} + \sqrt{144} = 19$
$$7 + 12 = 19$$

6.
(4)
$$-4^2 - 2^0 + 3^2(2x - 1) = 5^0(x - x^0) + 9$$
$$18x - 26 = x + 8$$
$$17x = 34$$
$$\mathbf{x = 2}$$

7.
(76)
(a) $3x + y + 4z = 15$

(b) $2x - 2y + 3z = -3$

(c) $x - 2z = 0$

(c') $x = 2z$

Substitute (c') into (a) and (b) to get:

(a') $y + 10z = 15$

(b') $-2y + 7z = -3$

2(a') $\quad 2y + 20z = 30$

(b') $\underline{-2y + 7z = -3}$

$\qquad\qquad 27z = 27$

$\qquad\qquad\quad z = 1$

(a') $y + 10 = 15$

$\qquad y = 5$

(c') $x = 2(1) = 2$

(2, 5, 1)

8.
(59)
(a) $\dfrac{1}{4}x - \dfrac{3}{2}y = -4$

$\qquad x - 6y = -16$

(a') $x = 6y - 16$

(b) $0.02x + 0.05y = 0.36$

(b') $2x + 5y = 36$

Substitute (a') into (b') to get:

(b'') $2(6y - 16) + 5y = 36$

$\qquad\quad 17y - 32 = 36$

$\qquad\qquad\qquad y = 4$

(a') $x = 6(4) - 16 = 8$

(8, 4)

9.
(26)
$$44^2 = Z^2 + 16^2$$
$$1936 = Z^2 + 256$$
$$1680 = Z^2$$
$$4\sqrt{105} = Z$$

$$\frac{77}{44} = \frac{Y + 4\sqrt{105}}{4\sqrt{105}}$$
$$308\sqrt{105} = 44Y + 176\sqrt{105}$$
$$132\sqrt{105} = 44Y$$
$$\mathbf{3\sqrt{105} = Y}$$

10.
(14)
$$y = \frac{2}{5}x + b$$
$$-3 = \frac{2}{5}(-2) + b$$
$$-\frac{11}{5} = b$$
$$y = \frac{2}{5}x - \frac{11}{5}$$

11.
(76)

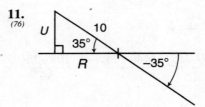

$R = 10 \cos 35° \approx 8.19$

$U = 10 \sin 35° \approx 5.74$

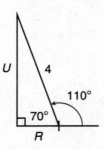

$R = 4 \cos 70° \approx 1.37$

$U = 4 \sin 70° \approx 3.76$

$\quad -8.19R + 5.74U$

$\underline{\quad -1.37R + 3.76U}$

$\mathbf{-9.56R + 9.50U}$

12.
(59)

$$\tan \theta = \frac{11}{6}$$

$$\theta \approx 61.39°$$

Subtract θ from 360° to get a fourth-quadrant angle:

$$360° - 61.39° \approx 298.61°$$

$$H = \sqrt{11^2 + 6^2} = \sqrt{157}$$

$$\sqrt{157}\underline{/298.61°}$$

13.
(B)

$$V = \frac{1}{3}A_{\text{Base}} \times H = \frac{1}{3}(\pi)(5)^2(12)$$

$$= 100\pi \approx \mathbf{314\ in.^3}$$

14.
(71)

(a) $\qquad ax^2 + bx + c = 0$

$$\left(x^2 + \frac{b}{a}x + \quad\right) = -\frac{c}{a}$$

$$x^2 + \frac{b}{a}x + \frac{b^2}{4ac} = \frac{b^2}{4a^2} - \frac{c}{a}$$

$$\left(x + \frac{b}{2a}\right)^2 = \frac{b^2 - 4ac}{4a^2}$$

$$x + \frac{b}{2a} = \pm\frac{\sqrt{b^2 - 4ac}}{2a}$$

$$x = \frac{-b \pm \sqrt{b^2 - 4ac}}{2a}$$

(b) $-9 + 4x = x^2$

$$0 = x^2 - 4x + 9$$

$$x = \frac{4 \pm \sqrt{(-4)^2 - 4(1)(9)}}{2(1)}$$

$$x = \frac{4 \pm \sqrt{-20}}{2} = \mathbf{2 \pm \sqrt{5}i}$$

15.
(75)

$$\frac{3x + 1}{x^2 - 16} + \frac{2x}{4 - x}$$

$$= \frac{3x + 1}{(x + 4)(x - 4)} - \frac{2x(x + 4)}{(x + 4)(x - 4)}$$

$$= \frac{\mathbf{-2x^2 - 5x + 1}}{\mathbf{x^2 - 16}}$$

16.
(46)

$$3\sqrt{7} + 5\sqrt{\frac{1}{7}} - 4\sqrt{28}$$

$$= 3\sqrt{7} + \frac{5}{\sqrt{7}} \cdot \frac{\sqrt{7}}{\sqrt{7}} - 8\sqrt{7}$$

$$= \frac{21\sqrt{7}}{7} + \frac{5\sqrt{7}}{7} - \frac{56\sqrt{7}}{7} = -\frac{\mathbf{30\sqrt{7}}}{\mathbf{7}}$$

17.
(73)

$$\frac{\sqrt{3} + 3}{\sqrt{3} - \sqrt{5}} \cdot \frac{\sqrt{3} + \sqrt{5}}{\sqrt{3} + \sqrt{5}}$$

$$= \frac{3 + 3\sqrt{3} + 3\sqrt{5} + \sqrt{15}}{3 - 5}$$

$$= -\frac{\mathbf{3 + 3\sqrt{3} + 3\sqrt{5} + \sqrt{15}}}{\mathbf{2}}$$

18.
(82)

$$\frac{1}{2 + \dfrac{a}{2 + \dfrac{2}{x}}} = \frac{1}{2 + \dfrac{a}{\dfrac{2x + 2}{x}}} = \frac{1}{2 + \dfrac{ax}{2x + 2}}$$

$$= \frac{1}{\dfrac{4x + 4 + ax}{2x + 2}} = \frac{\mathbf{2x + 2}}{\mathbf{4x + 4 + ax}}$$

19.
(81)

$$\frac{2 - 3i}{i - 5} \cdot \frac{i + 5}{i + 5} = \frac{2i + 10 - 3(-1) - 15i}{-1 - 25}$$

$$= \frac{13 - 13i}{-26} = -\frac{\mathbf{1}}{\mathbf{2}} + \frac{\mathbf{1}}{\mathbf{2}}i$$

20.
(70)

$$\frac{p}{m} = \frac{b}{m^2}\left(\frac{y}{a} + z\right)$$

$$\frac{mp}{b} = \frac{y}{a} + z$$

$$\frac{mp}{b} - z = \frac{y}{a}$$

$$\frac{\mathbf{amp}}{\mathbf{b}} - \mathbf{az} = \mathbf{y}$$

TEST 22

1.
(74)

$$\overset{D_B}{\underset{540}{\longmapsto}}$$

$$\overset{D_F}{\underset{135}{\longmapsto}}$$

$R_B T_B = 540;\ R_F T_F = 135$

$T_B = 3T_F;\ R_B = R_F + 15$

$3T_F(R_F + 15) = 540$

$3R_F T_F + 45T_F = 540$

$3(135) + 45T_F = 540$

$45T_F = 135$

$T_F = \textbf{3 hours}$

$T_B = 3(3) = \textbf{9 hours}$

$R_F(3) = 135$

$R_F = \textbf{45 mph}$

$R_B = 45 + 15 = \textbf{60 mph}$

2.
(5)

$f \times of = is$

$\dfrac{1}{6} \cdot 26{,}340 = D_O$

$D_O = \textbf{4390 ducks}$

3.
(61)

$0.4(30) + 0.8P_N = 0.5(30 + P_N)$

$12 + 0.8P_N = 15 + 0.5P_N$

$0.3P_N = 3$

$P_N = \textbf{10 gallons}$

4.
(83)

$\dfrac{x^{4a}(y^b)^{3a}\,x^{a/4}}{y^{ba/3}} = x^{4a}y^{3ab}x^{a/4}y^{-ab/3}$

$= \boldsymbol{x^{17a/4}y^{8ab/3}}$

5.
(83)

$\dfrac{(x^{a-3})^3}{x^{-3-2a}} = x^{3a-9}x^{3+2a} = \boldsymbol{x^{5a-6}}$

6.
(82)

$\dfrac{1}{x - \dfrac{a}{x - \dfrac{1}{a}}} = \dfrac{1}{x - \dfrac{a}{\dfrac{ax-1}{a}}} = \dfrac{1}{x - \dfrac{a^2}{ax-1}}$

$= \dfrac{1}{\dfrac{ax^2 - x - a^2}{ax - 1}} = \boldsymbol{\dfrac{ax - 1}{ax^2 - x - a^2}}$

7.
(66)

$\dfrac{2a - 3}{a - c} - \dfrac{3a - 1}{c - a} = \dfrac{2a - 3}{a - c} + \dfrac{3a - 1}{a - c}$

$= \boldsymbol{\dfrac{5a - 4}{a - c}}$

8.
(76)

(a) $x + 2y + 4z = 3$

(b) $x - y - 2z = -3$

(c) $3x + z = 0$

$$
\begin{array}{rl}
\text{(a)} & x + 2y + 4z = 3 \\
2\text{(b)} & \underline{2x - 2y - 4z = -6} \\
& 3x \qquad\qquad = -3 \\
& x = -1
\end{array}
$$

(c) $3(-1) + z = 0$

$z = 3$

(b) $-1 - y - 2(3) = -3$

$-y = 4$

$y = -4$

$\boldsymbol{(-1,\ -4,\ 3)}$

9.
(78)

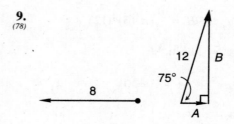

$A = 12 \cos 75° \approx 3.11$

$B = 12 \sin 75° \approx 11.59$

$$
\begin{array}{r}
3.11R + 11.59U \\
\underline{-8.00R + 0U} \\
\boldsymbol{-4.89R + 11.59U}
\end{array}
$$

$\tan \theta \approx \dfrac{11.59}{4.89}$

$\theta \approx 67.12°$

Subtract θ from $180°$ to get a second-quadrant angle:

$180° - 67.12° \approx 112.88°$

$F \approx \sqrt{(-4.89)^2 + 11.59^2} \approx 12.58$

$\boldsymbol{12.58\underline{/112.88°}}$

10.
(81)

$\dfrac{4 + 5i}{i - 2} \cdot \dfrac{i + 2}{i + 2} = \dfrac{4i - 5 + 8 + 10i}{-1 - 4}$

$= \dfrac{3 + 14i}{-5} = \boldsymbol{-\dfrac{3}{5} - \dfrac{14}{5}i}$

11.
(64)

$-\sqrt{-25} - \sqrt{2}\sqrt{-2} + \sqrt{-5}\sqrt{-5} + 3i - 2i^2 - 3i^3$

$= -5i - \sqrt{2}\sqrt{2}i - 5 + 3i + 2 - 3i(ii)$

$= -2i - 2i + 3i - 3 = \boldsymbol{-3 - i}$

12.
(46)
$$5\sqrt{\frac{3}{5}} + 2\sqrt{60} - 2\sqrt{\frac{5}{3}}$$

$$= \frac{5\sqrt{3}}{\sqrt{5}} \cdot \frac{\sqrt{5}}{\sqrt{5}} + 4\sqrt{15} - \frac{2\sqrt{5}}{\sqrt{3}} \cdot \frac{\sqrt{3}}{\sqrt{3}}$$

$$= \frac{5\sqrt{15}}{5} + 4\sqrt{15} - \frac{2\sqrt{15}}{3}$$

$$= \frac{3\sqrt{15}}{3} + \frac{12\sqrt{15}}{3} - \frac{2\sqrt{15}}{3} = \frac{13\sqrt{15}}{3}$$

13.
(73)
$$\frac{-5 - 3\sqrt{5}}{2 - 4\sqrt{5}} \cdot \frac{2 + 4\sqrt{5}}{2 + 4\sqrt{5}}$$

$$= \frac{-10 - 6\sqrt{5} - 20\sqrt{5} - 60}{4 - 80}$$

$$= \frac{-70 - 26\sqrt{5}}{-76} = \frac{35}{38} + \frac{13\sqrt{5}}{38}$$

14.
(70)
$$\frac{x + 1}{m} - b = c\left(\frac{y}{a} + x\right)$$

$$\frac{x + 1}{m} - b = \frac{cy}{a} + cx$$

$$ax + a - abm = cmy + acmx$$

$$a + ax - abm - acmx = cmy$$

$$a(1 + x - bm - cmx) = cmy$$

$$a = \frac{cmy}{1 + x - bm - cmx}$$

15. (a) $x^2 + y^2 = 25$
(85)

(b) $x - y = 1$

(b′) $x = 1 + y$

Substitute (b′) into (a) to get:

(a′) $(1 + y)^2 + y^2 = 25$

$1 + 2y + y^2 + y^2 = 25$

$2y^2 + 2y - 24 = 0$

$y^2 + y - 12 = 0$

$(y - 3)(y + 4) = 0$

$y = -4, 3$

Substitute these y-values into (b′) and solve for x:

(b′) $x = 1 + (-4) = -3$

(b′) $x = 1 + 3 = 4$

(-3, -4) and (4, 3)

16. $-x - 5 \ngeq -1$; $D = \{\text{Negative Integers}\}$
(86)

$-x - 5 < -1$

$-x < 4$

$x > -4$

17.
(47)
$$\frac{3 \text{ mi}^2}{\text{hr}} \times \frac{5280 \text{ ft}}{1 \text{ mi}} \times \frac{5280 \text{ ft}}{1 \text{ mi}} \times \frac{1 \text{ hr}}{60 \text{ min}}$$

$$= \frac{3(5280)(5280)}{60} \frac{\text{ft}^2}{\text{min}}$$

18.
(71)
$$2x^2 = 2x + 3$$

$$2x^2 - 2x - 3 = 0$$

$$x = \frac{2 \pm \sqrt{(-2)^2 - 4(2)(-3)}}{2(2)}$$

$$x = \frac{2 \pm \sqrt{28}}{4} = \frac{1}{2} \pm \frac{\sqrt{7}}{2}$$

19.
(62)
$$2x^2 + 2x = -3$$

$$\left(x^2 + x + \phantom{\frac{1}{4}}\right) = -\frac{3}{2}$$

$$x^2 + x + \frac{1}{4} = -\frac{3}{2} + \frac{1}{4}$$

$$\left(x + \frac{1}{2}\right)^2 = -\frac{5}{4}$$

$$x + \frac{1}{2} = \pm\frac{\sqrt{5}}{2}i$$

$$x = -\frac{1}{2} \pm \frac{\sqrt{5}}{2}i$$

20.
(54)
$$\frac{y}{a} = \frac{c}{b}$$

$$y = \frac{ac}{b}$$

TEST 23

1.
(29)

$R_B T_B + R_W T_W = 70$

$R_B = 20$; $R_W = 5$; $T_B + T_W = 5$

$20T_B + 5(5 - T_B) = 70$

$15T_B + 25 = 70$

$T_B = 3$

$T_W = 5 - 3 = 2$

$D_B = 20(3) = \mathbf{60 \text{ miles}}$

$D_W = 5(2) = \mathbf{10 \text{ miles}}$

2. $PV = nRT$
(88)

$$P = \frac{nRT}{V}$$

$$P = \frac{0.642(0.0821)(800)}{23} \approx \mathbf{1.83 \text{ atmospheres}}$$

3. (a) $N_S = 3N_F - 5$
(21)

(b) $10N_F = 5N_S - 5$

Substitute (a) into (b) to get:

(b′) $10N_F = 5(3N_F - 5) - 5$

$10N_F = 15N_F - 30$

$-5N_F = -30$

$N_F = \mathbf{6}$

(a) $N_S = 3(6) - 5 = \mathbf{13}$

4. $-1 \le x - 3 < 1; \; D = \{\text{Reals}\}$
(89)

$2 \le x < 4$

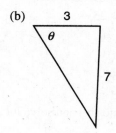

5. (a) $BT_D + 5T_D = 25$
(85)

(b) $\dfrac{BT_D - 5T_D = 15}{2BT_D \qquad = 40}$

(c) $BT_D = 20$

(a) $20 + 5T_D = 25$

$5T_D = 5$

$T_D = \mathbf{1}$

(c) $B(1) = 20$

$B = \mathbf{20}$

6. (a) $2x - y + z = -2$
(90)

(b) $x + 2y + 2z = 3$

(c) $2x - 2y + z = 0$

(a) $\quad 2x - y + z = -2$

$-1(c) \; \dfrac{-2x + 2y - z = 0}{y \qquad\qquad = -2}$

(b) $\quad x + 2y + 2z = 3$

$-2(a) \; \dfrac{-4x + 2y - 2z = 4}{-3x + 4y \qquad = 7}$

$-3x + 4(-2) = 7$

$-3x = 15$

$x = -5$

(c) $2(-5) - 2(-2) + z = 0$

$-6 + z = 0$

$z = 6$

$(-5, -2, 6)$

7.
(77)

$\sqrt{x} - 1 = \sqrt{x - 11}$

$x - 2\sqrt{x} + 1 = x - 11$

$-2\sqrt{x} = -12$

$4x = 144$

$x = 36$

Check: $\sqrt{36} - 1 = \sqrt{36 - 11}$

$5 = 5$

8. (a) $2(x + 2) = 3 \cdot 3$
(89)

$x + 2 = \dfrac{9}{2}$

$x = \dfrac{5}{2}$

(b) $4x = 2(6)$

$4x = 12$

$x = \mathbf{3}$

9. (a)
(76,59)

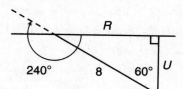

$R = 3 \cos 30° \approx 2.60$

$U = 3 \sin 30° = 1.50$

$R = 8 \cos 60° = 4$

$U = 8 \sin 60° \approx 6.93$

$2.60R + 1.50U$

$\dfrac{4.00R - 6.93U}{\mathbf{6.60R - 5.43U}}$

(b)

$\tan \theta = \dfrac{7}{3}$

$\theta \approx 66.80°$

Subtract θ from $360°$ to get a fourth-quadrant angle:

$360° - 66.80° \approx 293.20°$

$H = \sqrt{7^2 + 3^2} = \sqrt{58}$

$\mathbf{\sqrt{58} \, /293.20°}$

10. $D = \sqrt{(x_1 - x_2)^2 + (y_1 - y_2)^2}$
(88)
$\quad\; D = \sqrt{(-6 - 4)^2 + (2 - 6)^2}$

$\qquad = \sqrt{(-10)^2 + (-4)^2} = \sqrt{100 + 16}$

$\qquad = \sqrt{116} = \mathbf{2\sqrt{29}}$

11. $38 \text{ km}^2 \times \dfrac{1000 \text{ m}}{1 \text{ km}} \times \dfrac{1000 \text{ m}}{1 \text{ km}} \times \dfrac{100 \text{ cm}}{1 \text{ m}}$
(53)

$\quad \times \dfrac{100 \text{ cm}}{1 \text{ m}} \times \dfrac{1 \text{ in.}}{2.54 \text{ cm}} \times \dfrac{1 \text{ in.}}{2.54 \text{ cm}} \times \dfrac{1 \text{ ft}}{12 \text{ in.}}$

$\quad \times \dfrac{1 \text{ ft}}{12 \text{ in.}} \times \dfrac{1 \text{ mi}}{5280 \text{ ft}} \times \dfrac{1 \text{ mi}}{5280 \text{ ft}}$

$\quad = \dfrac{(38)(1000)(1000)(100)(100)}{(2.54)(2.54)(12)(12)(5280)(5280)} \text{ mi}^2$

12. $5 - 2x = 6x^2$
(71)
$\qquad 0 = 6x^2 + 2x - 5$

$\quad x = \dfrac{-2 \pm \sqrt{2^2 - 4(6)(-5)}}{2(6)}$

$\quad x = \dfrac{-2 \pm \sqrt{124}}{12} = -\dfrac{1}{6} \pm \dfrac{\sqrt{31}}{6}$

13.
(70)
$\qquad \dfrac{a}{c} - y = m\left(\dfrac{1}{p} + \dfrac{k}{x}\right)$

$\qquad \dfrac{a}{c} - y = \dfrac{m}{p} + \dfrac{mk}{x}$

$\qquad pxa - ypxc = mcx + kmcp$

$\quad xap - xcpy - xcm = ckmp$

$\quad x(ap - cpy - cm) = ckmp$

$\qquad\qquad x = \dfrac{\boldsymbol{ckmp}}{\boldsymbol{ap - cpy - cm}}$

14. Write the equation of the given line in slope-
(20) intercept form.

$\quad 2y - 5x = 8$

$\qquad 2y = 5x + 8$

$\qquad\; y = \dfrac{5}{2}x + 4$

$\qquad\; y = \dfrac{5}{2}x + b$

$\quad -2 = \dfrac{5}{2}(-3) + b$

$\quad \dfrac{11}{2} = b$

$\qquad\; y = \dfrac{5}{2}x + \dfrac{11}{2}$

15. $\dfrac{(a^3)^{x+y}a^{x-2y}c^{2x}}{c^{3x/2}}$
(83)

$\quad = a^{3x + 3y + x - 2y}c^{2x - 3x/2} = \mathbf{a^{4x+y}c^{x/2}}$

16. $\dfrac{p}{x + \dfrac{xp}{1 + \dfrac{p}{x}}} = \dfrac{p}{x + \dfrac{xp}{\dfrac{x + p}{x}}} = \dfrac{p}{x + \dfrac{x^2 p}{x + p}}$
(82)

$\quad = \dfrac{p}{\dfrac{x^2 + xp + x^2 p}{x + p}} = \dfrac{\boldsymbol{px + p^2}}{\boldsymbol{x^2 + xp + x^2 p}}$

17. $\dfrac{3 + 4i}{-2i - 5} \cdot \dfrac{-2i + 5}{-2i + 5} = \dfrac{-6i + 8 + 15 + 20i}{-4 - 25}$
(81)

$\quad = \dfrac{14i + 23}{-29} = -\dfrac{\mathbf{23}}{\mathbf{29}} - \dfrac{\mathbf{14}}{\mathbf{29}}\boldsymbol{i}$

18. $3i^2 + 4i - 2i^3 - \sqrt{-7}\sqrt{-7} + \sqrt{-3}\sqrt{-3}$
(64)
$\quad = 3(ii) + 4i - 2i(ii) + 7 - 3 = \mathbf{1 + 6i}$

19. $\dfrac{5 + \sqrt{5}}{8 - 2\sqrt{5}} \cdot \dfrac{8 + 2\sqrt{5}}{8 + 2\sqrt{5}}$
(73)

$\quad = \dfrac{40 + 8\sqrt{5} + 10\sqrt{5} + 10}{64 - 20} = \dfrac{50 + 18\sqrt{5}}{44}$

$\quad = \dfrac{\mathbf{25}}{\mathbf{22}} + \dfrac{\mathbf{9\sqrt{5}}}{\mathbf{22}}$

20. $\sqrt[3]{9\sqrt[5]{3}} = [3^2(3^{1/5})]^{1/3} = 3^{2/3}3^{1/15} = \mathbf{3^{11/15}}$
(47)

TEST 24

1. $\dfrac{P_1}{T_1} = \dfrac{P_2}{T_2}$
(57)

$\quad T_2 = \dfrac{P_2 T_1}{P_1}$

$\quad T_2 = \dfrac{300(80)}{400} = \mathbf{60 \text{ K}}$

2. $\quad A = kC$
(60)
$\quad 100 = k5$

$\quad\;\; 20 = k$

$\quad A = 20(30) = \mathbf{600 \text{ acorns}}$

3.
(92)

Downstream: $(B + W)T_D = D_D$ (a)

Upstream: $(B - W)T_U = D_U$ (b)

Since $T_D = T_U$ we use T_D in both equations

(a′) $BT_D + 10T_D = 140$

−1(b′) $\underline{-BT_D + 10T_D = -60}$

              $20T_D = \;\; 80$

                 $T_D = 4$

(a′) $B(4) + 10(4) = 140$

           $4B = 100$

           $B = \textbf{25 mph}$

4.
(94)

$g(x) = x^2 + 2$

$g(-1) = (-1)^2 + 2$

$g(-1) = 1 + 2$

$g(-1) = \textbf{3}$

5.
(93)

$5x = x^2 - 3$

$0 = x^2 - 5x - 3$

$b^2 - 4ac = (-5)^2 - 4(1)(-3) = 25 + 12 = 37$

$b^2 - 4ac > 0$

Two real number solutions

6.
(89)

(a) $x = \dfrac{175 - (360 - 175 - 110)}{2} = \dfrac{100}{2} = \textbf{50}$

(b) $4(4 + x) = 5^2$

      $16 + 4x = 25$

           $x = \dfrac{\textbf{9}}{\textbf{4}}$

7.
(71)

(a) $ax^2 + bx + c = 0$

$\left(x^2 + \dfrac{b}{a}x + \quad\right) = -\dfrac{c}{a}$

$x^2 + \dfrac{b}{a}x + \dfrac{b^2}{4a^2} = \dfrac{b^2}{4a^2} - \dfrac{c}{a}$

$\left(x + \dfrac{b}{2a}\right)^2 = \dfrac{b^2 - 4ac}{4a^2}$

$x + \dfrac{b}{2a} = \pm\dfrac{\sqrt{b^2 - 4ac}}{2a}$

$x = \dfrac{\boldsymbol{-b \pm \sqrt{b^2 - 4ac}}}{\boldsymbol{2a}}$

(b) $x = -2 - 3x^2$

$3x^2 + x + 2 = 0$

$x = \dfrac{-1 \pm \sqrt{1^2 - 4(3)(2)}}{2(3)}$

$x = \dfrac{-1 \pm \sqrt{-23}}{6} = -\dfrac{\textbf{1}}{\textbf{6}} \pm \dfrac{\sqrt{\textbf{23}}}{\textbf{6}}\boldsymbol{i}$

8.
(77)

$\sqrt{x} = \sqrt{x + 12} - 2$

$\sqrt{x} + 2 = \sqrt{x + 12}$

$x + 4\sqrt{x} + 4 = x + 12$

$4\sqrt{x} = 8$

$\sqrt{x} = 2$

$x = \textbf{4}$

Check: $\sqrt{4} = \sqrt{4 + 12} - 2$

        $2 = 2$

9.
(78)

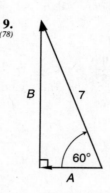

$A = 7 \cos 60° = 3.5$

$B = 7 \sin 60° \approx 6.06$

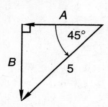

$A = 5 \cos 45° \approx 3.54$

$B = 5 \sin 45° \approx 3.54$

    $-3.50R + 6.06U$

    $\underline{-3.54R - 3.54U}$

    $\boldsymbol{-7.04R + 2.52U}$

$\tan \theta \approx \dfrac{2.52}{7.04}$

$\theta \approx 19.70°$

Subtract θ from 180° to get a second-quadrant angle:

$180° - 19.70° \approx 160.30°$

$F \approx \sqrt{(-7.04)^2 + 2.52^2} \approx 7.48$

7.48/160.30°

10.
(91)

(a) $2x + 6y \geq 18$

      $6y \geq -2x + 18$

        $y \geq -\dfrac{1}{3}x + 3$

(b) $-3x + y < -2$

$$y < 3x - 2$$

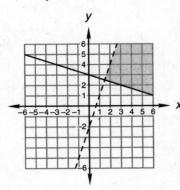

The region we wish to find is on or above the solid line and below the dashed line. This region is shaded in the figure above.

11.
(70)

$$\frac{x + z}{a} = a\left(\frac{b}{m} + \frac{1}{c}\right)$$

$$\frac{x + z}{a} = \frac{ab}{m} + \frac{a}{c}$$

$$xcm + cmz = a^2bc + a^2m$$

$$xcm = a^2bc + a^2m - cmz$$

$$x = \frac{a^2bc + a^2m - cmz}{cm}$$

12.
(90)

(a) $2x - y + z = 7$

(b) $x + y - z = -4$

(c) $x + 3y - 2z = -11$

(a) $2x - y + z = 7$

(b) $\underline{x + y - z = -4}$

    $3x \qquad\qquad = 3$

          $x = 1$

  $3(b)$ $3x + 3y - 3z = -12$

$-1(c)$ $\underline{-x - 3y + 2z = 11}$

   $2x \qquad\qquad -z = -1$

            $-z = -1 - 2(1)$

              $z = 3$

(b) $1 + y - 3 = -4$

        $y = -2$

(1, –2, 3)

13.
(85)

(a) $x^2 + y^2 = 5$

(b) $x - 3y = 5$

(b′) $x = 5 + 3y$

Substitute (b′) into (a) to get:

(a′) $(5 + 3y)^2 + y^2 = 5$

    $25 + 30y + 9y^2 + y^2 = 5$

       $10y^2 + 30y + 20 = 0$

        $(y + 2)(y + 1) = 0$

              $y = -2, -1$

Substitute these y-values into (b′) and solve for x:

(b′) $x = 5 + 3(-2) = -1$

(b′) $x = 5 + 3(-1) = 2$

(–1, –2) and **(2, –1)**

14.
(89)
$x + 2 \ngtr 1$ or $x - 3 \nleq -1$; $D = \{\text{Reals}\}$

$x + 2 \le 1$ or $x - 3 > -1$

    $x \le -1$ or $x > 2$

```
   ●———————————○——————
  -2 -1  0  1  2  3
```

15.
(83)
$\dfrac{m^{x/4}n^3}{m^{2x/3}(n^a)^3} = m^{x/4}n^3 m^{-2x/3}n^{-3a}$

$= m^{-5x/12}n^{3 - 3a}$

16.
(81)
$\dfrac{2 - 3i}{4 - i} \cdot \dfrac{4 + i}{4 + i} = \dfrac{8 - 12i + 2i + 3}{16 + 1}$

$= \dfrac{11 - 10i}{17} = \dfrac{\mathbf{11}}{\mathbf{17}} - \dfrac{\mathbf{10}}{\mathbf{17}}i$

17.
(46)
$5\sqrt{\dfrac{7}{20}} - 2\sqrt{\dfrac{20}{7}} - 3\sqrt{560}$

$= \dfrac{5\sqrt{7}}{2\sqrt{5}} \cdot \dfrac{\sqrt{5}}{\sqrt{5}} - \dfrac{4\sqrt{5}}{\sqrt{7}} \cdot \dfrac{\sqrt{7}}{\sqrt{7}} - 12\sqrt{35}$

$= \dfrac{5\sqrt{35}}{10} - \dfrac{4\sqrt{35}}{7} - 12\sqrt{35}$

$= \dfrac{7\sqrt{35}}{14} - \dfrac{8\sqrt{35}}{14} - \dfrac{168\sqrt{35}}{14} = -\dfrac{\mathbf{169\sqrt{35}}}{\mathbf{14}}$

18.
(73)
$\dfrac{5 + 2\sqrt{2}}{5\sqrt{2} - 2} \cdot \dfrac{5\sqrt{2} + 2}{5\sqrt{2} + 2}$

$= \dfrac{25\sqrt{2} + 20 + 10 + 4\sqrt{2}}{50 - 4} = \dfrac{30 + 29\sqrt{2}}{46}$

$= \dfrac{\mathbf{15}}{\mathbf{23}} + \dfrac{\mathbf{29\sqrt{2}}}{\mathbf{46}}$

19.
(82)
$\dfrac{a}{b + \dfrac{1}{1 + \dfrac{c}{x}}} = \dfrac{a}{b + \dfrac{1}{\dfrac{x + c}{x}}} = \dfrac{a}{b + \dfrac{x}{x + c}}$

$= \dfrac{a}{\dfrac{bx + bc + x}{x + c}} = \dfrac{\mathbf{ax + ac}}{\mathbf{bx + bc + x}}$

20.
(53)
$$\frac{1200 \text{ cm}^3}{\text{hr}} \times \frac{1 \text{ mL}}{1 \text{ cm}^3} \times \frac{1 \text{ L}}{1000 \text{ mL}} \times \frac{1 \text{ hr}}{60 \text{ min}}$$

$$\times \frac{1 \text{ min}}{60 \text{ s}} = \frac{1200}{(1000)(60)(60)} \frac{\text{L}}{\text{s}}$$

TEST 25

1. Variation method:
(96)

$$N_B = \frac{kN_G}{N_T}$$

$$24 = \frac{k(16)}{2}$$

$$k = 3$$

$$72 = \frac{3N_G}{6}$$

$$N_G = \textbf{144 girls}$$

Equal ratio method:

$$\frac{N_{B_1}}{N_{B_2}} = \frac{N_{G_1} N_{T_2}}{N_{G_2} N_{T_1}}$$

$$\frac{24}{72} = \frac{16(6)}{N_{G_2}(2)}$$

$$N_{G_2} = \textbf{144 girls}$$

2.
(74)

$$\overset{D_M}{\underset{888}{\longmapsto\!\!\!\!\longrightarrow}}$$

$$\overset{D_O}{\underset{280}{\longmapsto}}$$

$$R_M T_M = 888; \quad R_O T_O = 280$$

$$T_M = 3T_O; \quad R_M = R_O + 4$$

$$(R_O + 4)(3T_O) = 888$$

$$3R_O T_O + 12T_O = 888$$

$$840 + 12T_O = 888$$

$$12T_O = 48$$

$$T_O = \textbf{4 hours}$$

$$T_M = 3(4) = \textbf{12 hours}$$

$$R_O(4) = 280$$

$$R_O = \textbf{70 mph}$$

$$R_M(12) = 888$$

$$R_M = \textbf{74 mph}$$

3. Saline P_N + Saline D_N = Saline Total
(52)
$$(0.2)\, P_N + (0.5)D_N = 0.41(300)$$

(a) $0.2P_N + 0.5D_N = 123$

(b) $P_N + D_N = 300$

Substitute $P_N = 300 - D_N$ into (a) to get:

(a') $0.2(300 - D_N) + 0.5D_N = 123$

$$0.3D_N = 63$$

$$D_N = \textbf{210 gallons of}$$
$$\textbf{50\% solution}$$

(b) $P_N + 210 = 300$

$$P_N = \textbf{90 gallons of 20\% solution}$$

4. (a) $x - 4y = 12$
(95)

(a') $x = 4y + 12$

(b) $xy = 16$

Substitute (a') into (b) to get:

(b') $(4y + 12)y = 16$

$$4y^2 + 12y - 16 = 0$$

$$y^2 + 3y - 4 = 0$$

$$(y + 4)(y - 1) = 0$$

$$y = -4, 1$$

Substitute these y-values into (a') and solve for x.

(a') $x = 4(-4) + 12 = -4$

(a') $x = 4(1) + 12 = 16$

$(\textbf{-4, -4})$ and $(\textbf{16, 1})$

5. **(a)** and **(c)** are functions.
(94)

(b) is not a function because 3 has two images.

6. (a) $x + 2y > 6$
(91)

$$y > -\frac{1}{2}x + 3$$

(b) $y \geq 3$

The first step is to graph each of these lines.

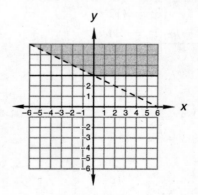

The region we wish to find is above the dashed line and on or above the solid line. This region is shaded in the figure above.

7.
(78)

$B = 12 \cos 30° \approx 10.39$

$C = 12 \sin 30° = 6$

$-10.39R - 6U$
$\underline{15.00R + 0U}$
$4.61R - 6U$

$\tan \theta \approx \dfrac{6}{4.61}$

$\theta \approx 52.46°$

Subtract θ from 360° to get a fourth-quadrant angle:

$360° - 52.46° \approx 307.54°$

$H \approx \sqrt{4.61^2 + (-6)^2} = \sqrt{57.25} \approx 7.57$

7.57 $\underline{/307.54°}$

8. $-4 \leq x + 1 < 3;\ D = \{\text{Reals}\}$
(89)
$-5 \leq x < 2$

9.
(70)

$\dfrac{x}{c + m} = p\left(\dfrac{e}{y} + \dfrac{f}{z}\right)$

$\dfrac{x}{c + m} = \dfrac{pe}{y} + \dfrac{pf}{z}$

$xyz = cpez + mpez + cpfy + mpfy$

$xyz - mpez - mpfy = c(pez + pfy)$

$c = \dfrac{xyz - mpez - mpfy}{pez + pfy}$

10. (a) $2x + 3y = 9$
(97)
(b) $3x + 4y = 12$

(b') $x = -\dfrac{4}{3}y + 4$

Substitute (b') into (a) to get:

(a') $2\left(-\dfrac{4}{3}y + 4\right) + 3y = 9$

$-\dfrac{8}{3}y + 8 + 3y = 9$

$\dfrac{1}{3}y = 1$

$y = 3$

(b') $x = -\dfrac{4}{3}(3) + 4 = 0$

(0, 3)

11. $N = \dfrac{3}{4} + \dfrac{2}{3}\left(3\dfrac{5}{12} - \dfrac{3}{4}\right) = \dfrac{3}{4} + \dfrac{2}{3}\left(\dfrac{41}{12} - \dfrac{9}{12}\right)$
(98)

$= \dfrac{3}{4} + \dfrac{2}{3}\left(\dfrac{32}{12}\right) = \dfrac{27}{36} + \dfrac{64}{36} = \dfrac{\mathbf{91}}{\mathbf{36}}$

12. $\dfrac{x - 1}{5} + 4 = \dfrac{x - 2}{3}$
(24)

$3x - 3 + 60 = 5x - 10$

$-2x = -67$

$x = \dfrac{\mathbf{67}}{\mathbf{2}}$

13. $-3x^2 + 2x - 7 = 0$
(62)

$\left(x^2 - \dfrac{2}{3}x + \right) = -\dfrac{7}{3}$

$x^2 - \dfrac{2}{3}x + \dfrac{1}{9} = -\dfrac{7}{3} + \dfrac{1}{9}$

$\left(x - \dfrac{1}{3}\right)^2 = -\dfrac{20}{9}$

$x - \dfrac{1}{3} = \pm\dfrac{2\sqrt{5}}{3}i$

$x = \dfrac{\mathbf{1}}{\mathbf{3}} \pm \dfrac{\mathbf{2\sqrt{5}}}{\mathbf{3}}i$

14. $\dfrac{5000 \text{ m}}{\text{hr}} \times \dfrac{100 \text{ cm}}{1 \text{ m}} \times \dfrac{1 \text{ in.}}{2.54 \text{ cm}} \times \dfrac{1 \text{ ft}}{12 \text{ in.}}$
(47)

$\times \dfrac{1 \text{ hr}}{60 \text{ min}} \times \dfrac{1 \text{ min}}{60 \text{ s}} = \dfrac{\mathbf{5000(100)}}{\mathbf{(2.54)(12)(60)(60)}} \dfrac{\mathbf{ft}}{\mathbf{s}}$

15. $\dfrac{(a^{3c})^{1/2}a^{3c}}{x^{c/3}} = a^{3c/2 + 3c}x^{-c/3} = \mathbf{a^{9c/2}x^{-c/3}}$
(83)

16. $\dfrac{5i^2 - 3i^3}{i^3 + 3i^2} = \dfrac{5(ii) - 3i(ii)}{i(ii) + 3(ii)}$
(81)

$= \dfrac{-5 + 3i}{-i - 3} \cdot \dfrac{-i + 3}{-i + 3} = \dfrac{5i + 3 - 15 + 9i}{-1 - 9}$

$= \dfrac{-12 + 14i}{-10} = \dfrac{\mathbf{6}}{\mathbf{5}} - \dfrac{\mathbf{7}}{\mathbf{5}}i$

17. $-(-\sqrt{-16}) - \sqrt{-5}\sqrt{-5} + 3 - 3i^3 - 3i^2 + 3i$
(64)

$= 4i + 5 + 3 + 3i + 3 + 3i = \mathbf{11 + 10i}$

18.
(82)
$$\cfrac{xy}{x + \cfrac{y}{x + \cfrac{1}{xy}}} = \cfrac{xy}{x + \cfrac{y}{\cfrac{x^2y + 1}{xy}}}$$

$$= \cfrac{xy}{x + \cfrac{xy^2}{x^2y + 1}} = \cfrac{xy}{\cfrac{x^3y + x + xy^2}{x^2y + 1}}$$

$$= \cfrac{x^3y^2 + xy}{x^3y + x + xy^2} = \boldsymbol{\cfrac{x^2y^2 + y}{x^2y + 1 + y^2}}$$

19.
(47)
$$\sqrt[5]{9\sqrt[4]{3}} = [3^2(3^{1/4})]^{1/5} = 3^{2/5}(3^{1/20}) = \boldsymbol{3^{9/20}}$$

20.
(46)
$$4\sqrt{\frac{5}{7}} + 3\sqrt{\frac{7}{5}} - 2\sqrt{315}$$

$$= \frac{4\sqrt{5}}{\sqrt{7}} \cdot \frac{\sqrt{7}}{\sqrt{7}} + \frac{3\sqrt{7}}{\sqrt{5}} \cdot \frac{\sqrt{5}}{\sqrt{5}} - 6\sqrt{35}$$

$$= \frac{4\sqrt{35}}{7} + \frac{3\sqrt{35}}{5} - 6\sqrt{35}$$

$$= \frac{20\sqrt{35}}{35} + \frac{21\sqrt{35}}{35} - \frac{210\sqrt{35}}{35} = \boldsymbol{-\frac{169\sqrt{35}}{35}}$$

TEST 26

1.
(101)
Purchase Price = Retail Price + Markup

          = 375 + 0.25(375)

          = **$468.75**

2.
(92)
Downstream: $(B + W)T_D = D_D$ (a)

Upstream: $(B - W)T_U = D_U$ (b)

Since $T_D = T_U$ we use T_D in both equations.

(a′) $15T_D + WT_D = 45$

(b′) $\dfrac{15T_D - WT_D = 30}{30T_D \qquad\quad = 75}$

             $T_D = 2.5$

(a′) $15(2.5) + W(2.5) = 45$

          $2.5W = 7.5$

            $W = \textbf{3 mph}$

3.
(55)
$N \quad N + 1 \quad N + 2$

   $2(N)(N + 1) = (N + 2)^2 + 4$

   $2N^2 + 2N = N^2 + 4N + 4 + 4$

  $N^2 - 2N - 8 = 0$

$(N - 4)(N + 2) = 0$

            $N = 4, -2$

The desired integers are **4, 5,** and **6** or **–2, –1,** and **0.**

4.
(99)
$|x| + 3 \not\geq 7; \; D = \{\text{Integers}\}$

$|x| + 3 \leq 7$

   $|x| \leq 4$

$x \geq -4$ and $x \leq 4$

5.
(98)
$$N = 2\frac{1}{3} + \frac{3}{4}\left(4\frac{1}{6} - 2\frac{1}{3}\right)$$

$$= \frac{7}{3} + \frac{3}{4}\left(\frac{25}{6} - \frac{14}{6}\right)$$

$$= \frac{7}{3} + \frac{3}{4}\left(\frac{11}{6}\right) = \frac{56}{24} + \frac{33}{24} = \boldsymbol{\frac{89}{24}}$$

6.
(97)
(a) $4x - 2y = -28$

(a′) $y = 2x + 14$

(b) $2x + 3y = -6$

Substitute (a′) into (b) to get:

(b′) $2x + 3(2x + 14) = -6$

          $8x + 42 = -6$

            $8x = -48$

             $x = -6$

(a′) $y = 2(-6) + 14$

    $y = 2$

(–6, 2)

7.
(95)
(a) $x^2 + y^2 = 36$

(a′) $y^2 = 36 - x^2$

(b) $2x^2 - y^2 = -9$

Substitute (a′) into (b) to get:

(b′) $2x^2 - (36 - x^2) = -9$

              $3x^2 = 27$

              $x^2 = 9$

               $x = \pm 3$

(a′) $y^2 = 36 - (3)^2$

    $y^2 = 27$

     $y = \pm 3\sqrt{3}$

(a′) $y^2 = 36 - (-3)^2$

    $y^2 = 27$

     $y = \pm 3\sqrt{3}$

$(3, \pm 3\sqrt{3})$ and $(-3, \pm\sqrt{3})$

8.
(90)

(a) $x - 3y - z = 7$

(b) $2x + y - 2z = 0$

(c) $x - y + z = 1$

2(a) $2x - 6y - 2z = 14$
−1(b) $\underline{-2x - y + 2z = 0}$
              $-7y = 14$
                  $y = -2$

(a) $x - 3y - z = 7$
(c) $\underline{x - y + z = 1}$
   $2x - 4y = 8$
   $2x - 4(-2) = 8$
      $2x = 0$
       $x = 0$

(c) $0 - (-2) + z = 1$
          $z = -1$

(0, −2, −1)

9.
(102)

$g(x) = x + 4; \ D = \{\text{Reals}\}$

$h(x) = x^2 - 6; \ D = \{\text{Integers}\}$

$g(x)h(x) = (x + 4)(x^2 - 6)$

$gh(x) = x^3 + 4x^2 - 6x - 24$

$gh(-4) = (-4)^3 + 4(-4)^2 - 6(-4) - 24$

$= -64 + 64 + 24 - 24 = \mathbf{0}$

10.
(91)

(a) $x \geq 2$

(b) $4x - 5y > 15$

$y < \dfrac{4}{5}x - 3$

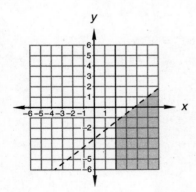

The region we wish to find is on or to the right of the solid line and below the dashed line. This region is shaded in the figure above.

11.
(100)

$y = -2x^2 + 8x - 6$

$-\dfrac{1}{2}y = (x^2 - 4x + 4) - 4 + 3$

$-\dfrac{1}{2}y = (x - 2)^2 - 1$

$y = -2(x - 2)^2 + 2$

From this we see:

(a) The graph opens downward.

(b) Axis of symmetry is $x = 2$.

(c) y-coordinate of vertex is 2.

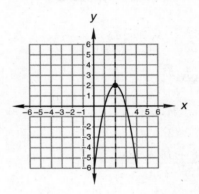

12.
(89)

$x - 2 \not> 4$ or $x - 1 > 7; \ D = \{\text{Integers}\}$

$x - 2 \leq 4$ or $x - 1 > 7$

$x \leq 6$ or $\quad\quad x > 8$

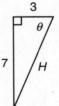

13.
(59)

$\tan \theta = \dfrac{7}{3}$

$\theta \approx 66.80°$

Add θ to 180° to get a third-quadrant angle:

$180° + 66.80° \approx 246.80°$

$H = \sqrt{(-3)^2 + (-7)^2} = \sqrt{58}$

$\underline{\mathbf{\sqrt{58} \, /246.80°}}$

14.
(55)
$$\frac{a}{c} = mn + \frac{a}{x + y}$$

$$ax + ay = cmnx + cmny + ac$$

$$ax + ay = c(mnx + mny + a)$$

$$\frac{ax + ay}{mnx + mny + a} = c$$

15.
(82)
$$\frac{cx}{1 - \dfrac{cx}{c - \dfrac{x}{c}}} = \frac{cx}{1 - \dfrac{cx}{\dfrac{c^2 - x}{c}}} = \frac{cx}{1 - \dfrac{c^2 x}{c^2 - x}}$$

$$= \frac{cx}{\dfrac{c^2 - x - c^2 x}{c^2 - x}} = \frac{c^3 x - cx^2}{c^2 - x - c^2 x}$$

16.
(83)
$$\frac{x^{c/4} y^{5c}}{x^{3c} y^{2c/3}} = x^{c/4} y^{5c} x^{-3c} y^{-2c/3}$$

$$= x^{-11c/4} y^{13c/3}$$

17.
(73)
$$\frac{4 - 3\sqrt{3}}{2\sqrt{3} - 4} \cdot \frac{2\sqrt{3} + 4}{2\sqrt{3} + 4}$$

$$= \frac{8\sqrt{3} - 18 + 16 - 12\sqrt{3}}{12 - 16}$$

$$= \frac{-2 - 4\sqrt{3}}{-4} = \frac{1}{2} + \sqrt{3}$$

18.
(46)
$$\sqrt[3]{x^5 y}\sqrt{x^2 y^4} = (x^5 y)^{1/3}(x^2 y^4)^{1/2}$$

$$= x^{5/3} y^{1/3} x y^2 = x^{8/3} y^{7/3}$$

19.
(81)
$$\frac{3i - 5}{-i^3 + 3i^2} = \frac{3i - 5}{i - 3} \cdot \frac{i + 3}{i + 3}$$

$$= \frac{-3 - 5i + 9i - 15}{-1 - 9}$$

$$= \frac{-18 + 4i}{-10} = \frac{9}{5} - \frac{2}{5}i$$

20.
(64)
$$3i^9 - \sqrt{-4}\sqrt{-4} + \sqrt{5}\sqrt{-5} - \sqrt{-9}$$

$$= 3i(ii)(ii)(ii)(ii) + 4 + \sqrt{5}\sqrt{5}i - 3i$$

$$= 3i + 4 + 5i - 3i = 4 + 5i$$

TEST 27

1.
(101)
Selling Price = Purchase Price + Markup

$$450 = 400 + \text{Markup}$$

$$\$50 = \text{Markup}$$

$$\% \text{ of } P_P = \frac{50}{400} \times 100\% = 12.5\%$$

2.
(74)

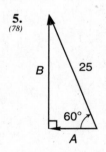

$$R_S T_S = 240 \qquad R_Y T_Y = 150$$

$$R_S = 2R_Y; \; T_S = T_Y - 1$$

$$2R_Y(T_Y - 1) = 240$$

$$2R_Y T_Y - 2R_Y = 240$$

$$2(150) - 2R_Y = 240$$

$$-2R_Y = -60$$

$$R_Y = \textbf{30 mph}$$

$$R_S = 2(30) = \textbf{60 mph}$$

$$30 T_Y = 150$$

$$T_Y = \textbf{5 hours}$$

$$T_S = 5 - 1 = \textbf{4 hours}$$

3.
(103)
$$
\begin{array}{r}
a^2 - ac + c^2 \\
a + c \overline{\smash{)}a^3 \qquad\qquad\qquad + c^3} \\
\underline{a^3 + a^2 c} \\
-a^2 c \\
\underline{-a^2 c - ac^2} \\
ac^2 + c^3 \\
\underline{ac^2 + c^3}
\end{array}
$$

4.
(94)
(b), **(c)**, and **(d)** are functions. (a) is not a function because 1 has two images.

5.
(78)

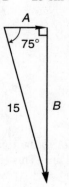

$$A = 25 \cos 60° = 12.5$$

$$B = 25 \sin 60° \approx 21.65$$

$A = 15 \cos 75° \approx 3.88$

$B = 15 \sin 75° \approx 14.49$

$$\begin{array}{r} -12.50R + 21.65U \\ 3.88R - 14.49U \\ \hline \mathbf{-8.62R + 7.16U} \end{array}$$

$\tan \theta \approx \dfrac{7.16}{8.62}$

$\quad \theta \approx 39.71°$

Subtract θ from $180°$ to get a second-quadrant angle:

$180° - 39.71° \approx 140.29°$

$F \approx \sqrt{(-8.62)^2 + (7.16)^2} \approx \sqrt{125.57} \approx 11.21$

11.21 $\underline{/140.29°}$

6.
(98)
$\begin{aligned} N &= 1\tfrac{1}{4} + \tfrac{1}{3}\left(3\tfrac{5}{12} - 1\tfrac{1}{4}\right) \\[2mm] &= \tfrac{5}{4} + \tfrac{1}{3}\left(\tfrac{41}{12} - \tfrac{5}{4}\right) \\[2mm] &= \tfrac{5}{4} + \tfrac{1}{3}\left(\tfrac{26}{12}\right) = \tfrac{45}{36} + \tfrac{26}{36} = \mathbf{\tfrac{71}{36}} \end{aligned}$

7.
(59)
(a) $\dfrac{2}{3}x - \dfrac{2}{5}y = 2$

$\qquad 10x - 6y = 30$

(a′) $y = \dfrac{5}{3}x - 5$

(b) $0.15x + 0.006y = 0.93$

(b′) $150x + 6y = 930$

Substitute (a′) into (b′) to get:

(b″) $150x + 6\left(\dfrac{5}{3}x - 5\right) = 930$

$\qquad\qquad\quad 160x - 30 = 930$

$\qquad\qquad\qquad\quad 160x = 960$

$\qquad\qquad\qquad\qquad x = 6$

(a′) $y = \dfrac{5}{3}(6) - 5 = 5$

(6, 5)

8.
(95)
(a) $5x + y = 8$

(a′) $y = -5x + 8$

(b) $xy = 3$

Substitute (a′) into (b) to get:

(b′) $\quad x(-5x + 8) = 3$

$\qquad -5x^2 + 8x - 3 = 0$

$x = \dfrac{-8 \pm \sqrt{8^2 - 4(-5)(-3)}}{2(-5)}$

$x = \dfrac{-8 \pm 2}{-10} = \dfrac{3}{5}, 1$

Substitute these x-values into (a′) and solve for y.

(a′) $y = -5\left(\dfrac{3}{5}\right) + 8 = 5$

(a′) $y = -5(1) + 8 = 3$

$\left(\dfrac{3}{5}, 5\right)$ and **(1, 3)**

9.
(100)
$y = -x^2 - 4x - 6$

$-y = (x^2 + 4x + 4) - 4 + 6$

$-y = (x + 2)^2 + 2$

$\quad y = -(x + 2)^2 - 2$

From this we see:

(a) The graph opens downward.

(b) Axis of symmetry is $x = -2$.

(c) y-coordinate of vertex is -2.

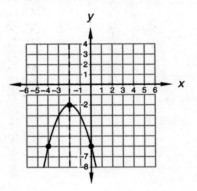

10.
(91)
(a) $3x + 2y < 8$

$\qquad y < -\dfrac{3}{2}x + 4$

(b) $y < -3$

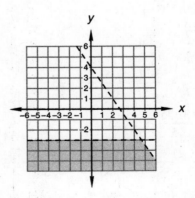

The region we wish to find is below both of the dashed lines. This region is shaded in the figure above.

11. $-|x| + 3 \not\geq 7$; $D = \{$Reals$\}$
(99)

$$-|x| + 3 < 7$$
$$-|x| < 4$$
$$|x| > -4$$

All real numbers

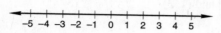

12. (a) $3x + 2y = 18$
(106)

(b) $2x - z = 4$

(c) $y + 2z = 21$

$$\begin{array}{r} \text{(a)}\quad 3x + 2y \qquad\;= 18 \\ -2\text{(c)} \quad\;\; -2y - 4z = -42 \\ \hline \text{(d)} \quad 3x \qquad -4z = -24 \end{array}$$

$$\begin{array}{r} \text{(d)}\quad 3x - 4z = -24 \\ -4\text{(b)} \quad -8x + 4z = -16 \\ \hline -5x \qquad\quad = -40 \\ x = 8 \end{array}$$

(b) $2(8) - z = 4$
$$z = 12$$

(c) $y + 2(12) = 21$
$$y = -3$$

(8, –3, 12)

13. $\sqrt{x - 12} = 6 - \sqrt{x}$
(77)
$$x - 12 = 36 - 12\sqrt{x} + x$$
$$-48 = -12\sqrt{x}$$
$$4 = \sqrt{x}$$
$$\mathbf{16 = x}$$

Check: $\sqrt{16 - 12} = 6 - \sqrt{16}$
$$2 = 2$$

14. $2\sqrt{\dfrac{5}{7}} - 3\sqrt{\dfrac{7}{5}} + \sqrt{140}$
(46)

$$= \frac{2\sqrt{5}}{\sqrt{7}} \cdot \frac{\sqrt{7}}{\sqrt{7}} - \frac{3\sqrt{7}}{\sqrt{5}} \cdot \frac{\sqrt{5}}{\sqrt{5}} + 2\sqrt{35}$$

$$= \frac{2\sqrt{35}}{7} - \frac{3\sqrt{35}}{5} + 2\sqrt{35}$$

$$= \frac{10\sqrt{35}}{35} - \frac{21\sqrt{35}}{35} + \frac{70\sqrt{35}}{35} = \mathbf{\frac{59\sqrt{35}}{35}}$$

15. $\dfrac{a^{2m}(a^{m/2 + 1})^4 c^n}{c^{n/2} a^{m/3}} = a^{2m}(a^{2m + 4})c^n c^{-n/2} a^{-m/3}$
(83)

$$= a^{11m/3 + 4} c^{n/2}$$

16. $\dfrac{3 - 2i - 4i^3}{i + 2i^2 - 5i^3}$
(81)

$$= \frac{3 + 2i}{6i - 2} \cdot \frac{6i + 2}{6i + 2}$$

$$= \frac{18i - 12 + 6 + 4i}{-36 - 4}$$

$$= \frac{-6 + 22i}{-40} = \mathbf{\frac{3}{20} - \frac{11}{20}i}$$

17. (a) Since this is a 30°-60°-90° triangle:
(66,79)

$$1 \times \overrightarrow{SF} = 6$$
$$\overrightarrow{SF} = 6$$
$$W = 6(\sqrt{3}) = \mathbf{6\sqrt{3}}$$
$$x = 6(2) = \mathbf{12}$$

(b) Since this is a 45°-45°-90° triangle:

$$1 \times \overrightarrow{SF} = 7$$
$$\overrightarrow{SF} = 7$$
$$y = \sqrt{2}(7) = \mathbf{7\sqrt{2}}$$
$$z = 1(7) = \mathbf{7}$$

18. (a) $N = 0.00601\ 01\ 01\ \ldots$
(104)

$$\begin{array}{r} 1000N = 6.01\ 01\ 01\ \ldots \\ 10N = 0.06\ 01\ 01\ \ldots \\ \hline 990N = 5.95 \end{array}$$

$$N = \frac{595}{99,000} = \mathbf{\frac{119}{19,800}}$$

(b) $N = 0.032\ 32\ 32\ \ldots$

$$\begin{array}{r} 100N = 3.2\ 32\ 32\ \ldots \\ N = 0.0\ 32\ 32\ \ldots \\ \hline 99N = 3.2 \end{array}$$

$$N = \frac{32}{990} = \mathbf{\frac{16}{495}}$$

19.
(105)
$$3x^2 + 5x - 2 = 0$$
$$3x^2 + 6x - x - 2 = 0$$
$$3x(x + 2) - (x + 2) = 0$$
$$(3x - 1)(x + 2) = 0$$
$$3x - 1 = 0 \qquad x + 2 = 0$$
$$x = \frac{1}{3} \qquad\quad x = -2$$

$$\mathbf{\frac{1}{3},\ -2}$$

20.
(105)

$$3x^2 - 13x + 12 = 0$$
$$3x^2 - 9x - 4x + 12 = 0$$
$$3x(x - 3) - 4(x - 3) = 0$$
$$(3x - 4)(x - 3) = 0$$
$$3x - 4 = 0 \qquad x - 3 = 0$$
$$x = \frac{4}{3} \qquad\qquad x = 3$$

$$\frac{4}{3}, 3$$

TEST 28

1. Downstream: $(B + W)T_D = D_D$ (a)
(92)

Upstream: $(B - W)T_U = D_U$ (b)

(a′) $4B + 4W = 112$

(b′) $6B - 6W = 48$

3(a′) $12B + 12W = 336$

2(b′) $\underline{12B - 12W = \ \ 96}$

$\qquad 24B \qquad\quad = 432$

$\qquad\qquad\qquad B = \textbf{18 mph}$

(b′) $6(18) - 6W = 48$

$\qquad\qquad -6W = -60$

$\qquad\qquad\quad W = \textbf{10 mph}$

2. Calcium: $1 \times 40 = 40$
(37)

Sulfur: $1 \times 32 = 32$

Oxygen: $4 \times 16 = 64$

Total: $\qquad\quad = 136$

$$\frac{64}{136} = \frac{O}{1224}$$

$$136(O) = 64(1224)$$

$$O = \textbf{576 g}$$

3. $T = $ tens digit
(107)

$U = $ units digit

$10T + U = $ original number

$10U + T = $ reversed number

(a) $T + U = 15$

(a′) $T = 15 - U$

(b) $10U + T = 10T + U + 9$

Substitute (a′) into (b) to get:

(b′) $10U + (15 - U) = 10(15 - U) + U + 9$

$\qquad\qquad 9U + 15 = -9U + 159$

$\qquad\qquad\qquad 18U = 144$

$\qquad\qquad\qquad\quad U = 8$

(a′) $T = 15 - (8) = 7$

Original number = **78**

4. Selling Price = Purchase Price + Markup
(101)

$\qquad 30{,}000 = 25{,}000 + $ Markup

$\qquad \$5000 = $ Markup

% of Purchase Price $= \dfrac{5000}{25{,}000} \times 100\%$

$= \textbf{20\% of Purchase Price}$

% of Selling Price $= \dfrac{5000}{30{,}000} \times 100\%$

$= \textbf{16.67\% of Selling Price}$

5. (a) $x + 2y + z = 7$
(90)

(b) $3x - y + z = -12$

(c) $4x + 3y - 2z = 9$

$\quad$ (a) $\quad x + 2y + z = \ 7$

-1(b) $\underline{-3x + \ y - z = 12}$

$\quad$ (d) $-2x + 3y \qquad = 19$

2(a) $2x + 4y + 2z = 14$

$\ $ (c) $\underline{4x + 3y - 2z = \ 9}$

$\ $ (e) $6x + 7y \qquad = 23$

3(d) $-6x + \ 9y = 57$

$\ $ (e) $\underline{\ 6x + \ 7y = 23}$

$\qquad\qquad 16y = 80$

$\qquad\qquad\ y = 5$

(d) $-2x + 3(5) = 19$

$\qquad\quad -2x = 4$

$\qquad\qquad\ x = -2$

(a) $-2 + 2(5) + z = 7$

$\qquad\qquad\qquad z = -1$

(−2, 5, −1)

6. (a) $\dfrac{2}{3}x - \dfrac{2}{5}y = 2$
(59)

(a′) $10x - 6y = 30$

(b) $0.03x + 0.04y = 0.67$

(b′) $3x + 4y = 67$

2(a′) $20x - 12y = \ \ 60$

3(b′) $\underline{\ 9x + 12y = 201}$

$\qquad 29x \qquad\quad = 261$

$\qquad\qquad\quad x = 9$

(b′) $3(9) + 4y = 67$

$\qquad\qquad 4y = 40$

$\qquad\qquad\ y = 10$

(9, 10)

7. $N = 3.6123\ 123\ 123\ \dots$
(104)

$1000N = 3612.3123\ 123\ 123\ \dots$

$\underline{\quad N = \quad\ \ 3.6123\ 123\ 123\ \dots}$

$999N = 3608.7$

$$N = \frac{36{,}087}{9990} = \mathbf{\frac{12{,}029}{3330}}$$

8. $y = -x^2 + 4x + 1$
(100)

$-y = (x^2 - 4x + 4) - 4 - 1$

$-y = (x - 2)^2 - 5$

$y = -(x - 2)^2 + 5$

From this we see:

(a) The graph opens downward.

(b) Axis of symmetry is $x = 2$.

(c) y-coordinate of vertex is 5.

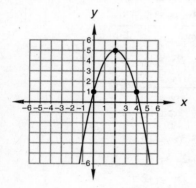

9. (a) $4x < -12$
(91)

$\qquad x < -3$

(b) $2x + y \geq -4$

$\qquad\ \ y \geq -2x - 4$

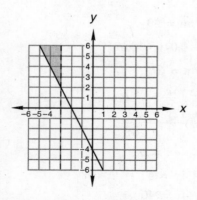

The region we wish to find is to the left of the dashed line and on or above the solid line. This region is shaded in the figure above.

10. $\qquad\qquad 14x^3 = 42x - 7x^2$
(105)

$\qquad\ 7x(2x^2 + x - 6) = 0$

$7x(2x^2 - 3x + 4x - 6) = 0$

$7x[x(2x - 3) + 2(2x - 3)] = 0$

$\qquad\ 7x(x + 2)(2x - 3) = 0$

$7x = 0 \qquad x + 2 = 0 \qquad 2x - 3 = 0$

$x = 0 \qquad\quad x = -2 \qquad\quad x = \dfrac{3}{2}$

$\mathbf{0,\ -2,\ \dfrac{3}{2}}$

11. $\qquad\ \sqrt{x} + 2 = \sqrt{x + 12}$
(77)

$x + 4\sqrt{x} + 4 = x + 12$

$\qquad\quad 4\sqrt{x} = 8$

$\qquad\quad\ \sqrt{x} = 2$

$\qquad\qquad x = \mathbf{4}$

Check: $\sqrt{4} + 2 = \sqrt{4 + 12}$

$\qquad\qquad 4 = 4$

12. $27x^3y^6 - a^9c^{12} = (3xy^2)^3 - (a^3c^4)^3$
(108)
$\mathbf{= (3xy^2 - a^3c^4)(9x^2y^4 + 3a^3c^4xy^2 + a^6c^8)}$

13. $a^{3/2} + c^{1/4}$
(109)
$\dfrac{a^{3/2} + c^{1/4}}{a^3 \ + \ a^{3/2}c^{1/4}}$

$\dfrac{\qquad\quad a^{3/2}c^{1/4} + c^{1/2}}{a^3 + 2a^{3/2}c^{1/4} + c^{1/2}}$

14. $f(x) = (x - 1)^2;\ D = \{\text{Reals}\}$
(102)

$g(x) = x + 3;\ D = \{\text{Integers}\}$

$f(x)g(x) = (x - 1)^2(x + 3)$

$\quad fg(x) = (x - 1)^2(x + 3)$

$\quad fg(2) = (2 - 1)^2(2 + 3) = 1^2(5) = \mathbf{5}$

15. $x^2 - 5x + 6 \geq 0;\ D = \{\text{Reals}\}$
(110)

$(x - 3)(x - 2) \geq 0$

$(\text{Pos})(\text{Pos}) \geq 0$

$x - 3 \geq 0$ and $x - 2 \geq 0$

$\quad x \geq 3$ and $\qquad x \geq 2$

$(\text{Neg})(\text{Neg}) \geq 0$

$x - 3 \leq 0$ and $x - 2 \leq 0$

$\quad x \leq 3$ and $\qquad x \leq 2$

Thus, the solution is $\mathbf{x \leq 2}$ or $\mathbf{x \geq 3}$

16. $\dfrac{400 \text{ L}}{\text{min}} \times \dfrac{1000 \text{ mL}}{\text{L}} \times \dfrac{1 \text{ min}}{60 \text{ s}}$
(47)

$= \dfrac{400(1000)}{60} \dfrac{\text{mL}}{\text{s}}$

17. (a) $x + 3y = 7$
(95)

(a') $x = 7 - 3y$

(b) $xy = 2$

Substitute (a') into (b) to get:

(b') $(7 - 3y)y = 2$

$7y - 3y^2 = 2$

$0 = 3y^2 - 7y + 2$

$0 = 3y^2 - y - 6y + 2$

$0 = y(3y - 1) - 2(3y - 1)$

$0 = (y - 2)(3y - 1)$

$y - 2 = 0 \qquad 3y - 1 = 0$

$y = 2 \qquad y = \dfrac{1}{3}$

Substitute these y-values into (a') and solve for x.

(a') $x = 7 - 3(2) = 1$

(a') $x = 7 - 3\left(\dfrac{1}{3}\right) = 6$

$(1, 2)$ and $\left(6, \dfrac{1}{3}\right)$

18. $\dfrac{4i - 3i^2 - 2}{\sqrt{-25} - \sqrt{-3}\sqrt{-3}} = \dfrac{4i + 1}{5i + 3} \cdot \dfrac{5i - 3}{5i - 3}$
(81)

$= \dfrac{-20 + 5i - 12i - 3}{-34}$

$= \dfrac{-23 - 7i}{-34} = \dfrac{23}{34} + \dfrac{7}{34}i$

19. $\sqrt[5]{16\sqrt{2}} = \left[2^4(2^{1/2})\right]^{1/5}$
(47)
$= 2^{4/5}2^{1/10} = 2^{9/10}$

20.
(59)

$\tan \theta = \dfrac{17}{8}$

$\theta \approx 64.80°$

Subtract θ from $180°$ to get a second-quadrant angle:

$180° - 64.80° \approx 115.20°$

$H = \sqrt{17^2 + 8^2} = \sqrt{353} \approx 18.79$

18.79 $\underline{/115.20°}$

TEST 29

1.
(74)

$R_A T_A = 48 \qquad R_B T_B = 30$

$R_A = 4R_B; \; T_A = T_B - 3$

$4R_B(T_B - 3) = 48$

$4R_B T_B - 12R_B = 48$

$4(30) - 12R_B = 48$

$-12R_B = -72$

$R_B = \textbf{6 mph}$

$R_A = 4(6) = \textbf{24 mph}$

$6T_B = 30$

$T_B = \textbf{5 hours}$

$T_A = 5 - 3 = \textbf{2 hours}$

2. (a) $N_N + N_D + N_Q = 27$
(111)
(b) $5N_N + 10N_D + 25N_Q = 300$

(c) $N_N = 3N_D$

$\begin{array}{ll} 25\text{(a)} & 25N_N + 25N_D + 25N_Q = 675 \\ -1\text{(b)} & -5N_N - 10N_D - 25N_Q = -300 \\ \hline & 20N_N + 15N_D = 375 \end{array}$

(d) $4N_N + 3N_D = 75$

Substitute (c) into (d) to get:

(d') $4(3N_D) + 3N_D = 75$

$15N_D = 75$

$N_D = \textbf{5 dimes}$

(c) $N_N = 3(5) = \textbf{15 nickels}$

(a) $15 + 5 + N_Q = 27$

$N_Q = \textbf{7 quarters}$

3.
(107)

T = tens digit

U = units digit

$10T + U$ = original number

$10U + T$ = reversed number

(a) $T + U = 10$

(a′) $T = 10 - U$

(b) $10U + T = 3(10T + U) - 2$

    $10U + T = 30T + 3U - 2$

(b′) $7U = 29T - 2$

Substitute (a′) into (b′) to get:

(b″) $7U = 29(10 - U) - 2$

       $7U = 288 - 29U$

     $36U = 288$

       $U = 8$

(a′) $T = 10 - 8 = 2$

Original number = **28**

4. $x^2 + x - 6 \leq 0$; D = {Reals}
(112)

$(x + 3)(x - 2) \leq 0$

$(\text{Pos})(\text{Neg}) \leq 0$

$x + 3 \geq 0$ and $x - 2 \leq 0$

    $x \geq -3$ and $x \leq 2$

$(\text{Neg})(\text{Pos}) \leq 0$

$x + 3 \leq 0$ and $x - 2 \geq 0$

    $x \leq -3$ and $x \geq 2$

There are no real numbers that satisfy the second conjunction, so the solution must be **−3 ≤ x ≤ 2.**

```
  ←—+——●——+——+——+——●——+——→
    -4 -3 -2 -1  0  1  2  3
```

5. $x^2 + 7x + 10 < 0$; D = {Reals}
(112)

$(x + 5)(x + 2) < 0$

$(\text{Pos})(\text{Neg}) < 0$

$x + 5 > 0$ and $x + 2 < 0$

    $x > -5$ and $x < -2$

$(\text{Neg})(\text{Pos}) < 0$

$x + 5 < 0$ and $x + 2 > 0$

    $x < -5$ and $x > -2$

There are no real numbers that satisfy the second conjunction, so the solution must be **−5 < x < −2.**

```
  ←—+——○——+——+——○——+——→
    -6 -5 -4 -3 -2 -1
```

6. $|x| + 3 \not> 8$; D = {Reals}
(99)

    $|x| \leq 5$

$x \leq 5$ and $x \geq -5$

```
  ←—+——●——+——+——+——+——+——+——+——+——+——●——+——→
    -6 -5 -4 -3 -2 -1  0  1  2  3  4  5  6
```

7.
(109)

$$x^{1/4} + y^{1/2}$$

$$\frac{x^{-1/2} - y^{-1/4}}{x^{-1/4} + x^{-1/2}y^{1/2}}$$

$$\frac{-x^{1/4}y^{-1/4} - y^{1/4}}{x^{-1/4} + x^{-1/2}y^{1/2} - x^{1/4}y^{-1/4} - y^{1/4}}$$

8. $8a^3c^9 - 64x^{12}y^6 = (2ac^3)^3 - (4x^4y^2)^3$
(108)

$= (2ac^3 - 4x^4y^2)(4a^2c^6 + 8ac^3x^4y^2 + 16x^8y^4)$

9. (a) $x \approx 0.51$
(113)

(b) $x = e^{6.417} \approx 612.16$

(c) $x \approx 2.66$

(d) $x = 10^{2.735} \approx 543.25$

10. $N = 0.00320\ 20\ 20\ \ldots$
(104)

$1000N = 3.20\ 20\ 20\ \ldots$

$\underline{10N = 0.03\ 20\ 20\ \ldots}$

$990N = 3.17$

          $N = \dfrac{317}{99,000}$

11. $y = x^2 + 4x - 2$
(100)

$y = (x^2 + 4x + 4) - 4 - 2$

$y = (x + 2)^2 - 6$

From this we see:

(a) The graph opens upward.

(b) Axis of symmetry is $x = -2$.

(c) y-coordinate of vertex is -6.

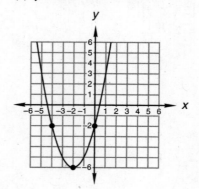

12. **Region B,** since this region includes all points that
(114) are either inside or on the circle and above the line

13. (a) $1\frac{1}{5}x + \frac{1}{2}y = 22$
(59)

(a') $12x + 5y = 220$

(b) $0.1x + 0.5y = 11$

(b') $x + 5y = 110$

 (a') $12x + 5y = 220$

$-1(b')$ $\underline{-x - 5y = -110}$

$11x = 110$

$x = 10$

(b') $10 + 5y = 110$

$y = 20$

(10, 20)

14. (a) $4x + 2y = 10$
(106)

(b) $-2x + z = -1$

(c) $x + 3z = 11$

 (b) $-2x + z = -1$

$2(c)$ $\underline{2x + 6z = 22}$

$7z = 21$

$z = 3$

(c) $x + 3(3) = 11$

$x = 2$

(a) $4(2) + 2y = 10$

$2y = 2$

$y = 1$

(2, 1, 3)

15. (a) $x^2 + y^2 = 34$
(85)

(b) $2x - y = 1$

(b') $y = 2x - 1$

Substitute (b') into (a) to get:

(a) $x^2 + (2x - 1)^2 = 34$

$x^2 + 4x^2 - 4x + 1 = 34$

$5x^2 - 4x - 33 = 0$

$5x^2 + 11x - 15x - 33 = 0$

$x(5x + 11) - 3(5x + 11) = 0$

$(x - 3)(5x + 11) = 0$

$x - 3 = 0 5x + 11 = 0$

$x = 3 x = -\dfrac{11}{5}$

Substitute these x-values into (b') and solve for y.

(b') $y = 2(3) - 1 = 5$

(b') $y = 2\left(-\dfrac{11}{5}\right) - 1 = -\dfrac{27}{5}$

(3, 5) and $\left(-\dfrac{11}{5}, -\dfrac{27}{5}\right)$

16. $N = 2\dfrac{5}{6} + \dfrac{3}{4}\left(8\dfrac{5}{12} - 2\dfrac{5}{6}\right)$
(98)

$ = \dfrac{17}{6} + \dfrac{3}{4}\left(\dfrac{101}{12} - \dfrac{17}{6}\right)$

$ = \dfrac{17}{6} + \dfrac{3}{4}\left(\dfrac{67}{12}\right) = \dfrac{136}{48} + \dfrac{201}{48} = \dfrac{\mathbf{337}}{\mathbf{48}}$

17. $\dfrac{4i^2 - 3i}{-\sqrt{-5}\sqrt{-5} + 2i^3} = \dfrac{-4 - 3i}{5 - 2i} \cdot \dfrac{5 + 2i}{5 + 2i}$
(81)

$ = \dfrac{-20 - 15i - 8i + 6}{25 + 4}$

$ = \dfrac{-14 - 23i}{29} = -\dfrac{\mathbf{14}}{\mathbf{29}} - \dfrac{\mathbf{23}}{\mathbf{29}}i$

18. $\dfrac{\sqrt{2} - 5}{5\sqrt{2} + 3} \cdot \dfrac{5\sqrt{2} - 3}{5\sqrt{2} - 3}$
(73)

$ = \dfrac{10 - 25\sqrt{2} - 3\sqrt{2} + 15}{50 - 9}$

$ = \dfrac{25 - 28\sqrt{2}}{41} = \dfrac{\mathbf{25}}{\mathbf{41}} - \dfrac{\mathbf{28}}{\mathbf{41}}\sqrt{2}$

19. $3x = 2x^2 - 2$
(105)

$0 = 2x^2 - 3x - 2$

$0 = 2x^2 - 4x + x - 2$

$ = 2x(x - 2) + (x - 2)$

$ = (2x + 1)(x - 2)$

$2x + 1 = 0 x - 2 = 0$

$x = -\dfrac{1}{2} x = 2$

$-\dfrac{\mathbf{1}}{\mathbf{2}}, \mathbf{2}$

20. $-4x = 2x^2 - 7$
(71)

$0 = 2x^2 + 4x - 7$

$x = \dfrac{-4 \pm \sqrt{4^2 - 4(2)(-7)}}{2(2)}$

$ = \dfrac{-4 \pm 6\sqrt{2}}{4} = \mathbf{-1 \pm \dfrac{3\sqrt{2}}{2}}$

TEST 30

1.
(92)
Downstream: $(B + W)T_D = D_D$ (a)

Upstream: $(B - W)T_U = D_U$ (b)

Since $T_U = 2T_D$ substitute to get:

(a′) $BT_D + 3T_D = 24$

(b′) $2BT_D - 6T_D = 12$

 2(a′) $\quad 2BT_D + 6T_D = 48$
-1(b′) $\quad \underline{-2BT_D + 6T_D = -12}$
$ 12T_D = 36$
$ T_D = 3$

(a′) $\quad B(3) + 3(3) = 24$
$ 3B = 15$
$ \mathbf{B = 5\ mph}$

2.
(111)
(a) $N_N + N_D + N_Q = 48$

(b) $5N_N + 10N_D + 25N_Q = 825$

(c) $N_Q = 2N_D$

 5(a) $\quad 5N_N + 5N_D + 5N_Q = 240$
-1(b) $\quad \underline{-5N_N - 10N_D - 25N_Q = -825}$
 (d) $ -5N_D - 20N_Q = -585$

Substitute (c) into (d) to get:

(d′) $-5N_D - 20(2N_D) = -585$
$ -45N_D = -585$
$ N_D = \mathbf{13\ dimes}$

(c) $N_Q = 2(13) = \mathbf{26\ quarters}$

(a) $N_N + 13 + 26 = 48$
$ N_N = \mathbf{9\ nickels}$

3.
(55)
$5N \qquad 5(N + 1) \qquad 5(N + 2)$

$(5N)^2 + [5(N + 1)]^2 = [5(N + 2)]^2 + 125$

$25N^2 + 25N^2 + 50N + 25$
$\qquad = 25N^2 + 100N + 100 + 125$

$25N^2 - 50N - 200 = 0$
$\qquad N^2 - 2N - 8 = 0$
$\qquad (N - 4)(N + 2) = 0$
$\qquad\qquad N = 4, -2$

The desired integers are **–10, –5,** and **0** or **20, 25,** and **30.**

4.
(113)
$\dfrac{e^{24.71}e^{-1.46}}{e^{11.04}} = e^{24.71\ -\ 1.46\ -\ 11.04}$

$= e^{12.21} \approx \mathbf{2.01 \times 10^5}$

5.
(113)
$\text{pH} = -\log \text{H}^+$

$\text{pH} = -\log (2.68 \times 10^{-3})$

$\text{pH} \approx \mathbf{2.57}$

6.
(115)
$A_t = Pe^{rt}$

$A_9 = 13{,}000e^{0.085(9)}$

$A_9 \approx \mathbf{\$27{,}937}$

7.
(13)

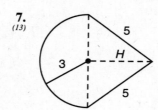

$5^2 = H^2 + 3^2$

$25 = H^2 + 9$

$16 = H^2$

$4 = H$

$A_{\text{Base}} = \dfrac{1}{2}(\pi)(3)^2 + \dfrac{1}{2}(6)(4)$

$\qquad = \dfrac{9}{2}\pi + 12 \approx 26.13\ \text{m}^2$

$V = A_{\text{Base}} \times \text{height}$
$\quad \approx 26.13(3) \approx \mathbf{78.39\ m^3}$

8.
(117)
$x^2 - 6x + 5 \le 0;\ D = \{\text{Reals}\}$

$(x - 5)(x - 1) \le 0$

$(\text{Pos})(\text{Neg}) \le 0$

$x - 5 \ge 0$ and $x - 1 \le 0$

$\qquad x \ge 5$ and $\qquad x \le 1$

$(\text{Neg})(\text{Pos}) \le 0$

$x - 5 \le 0$ and $x - 1 \ge 0$

$\qquad x \le 5$ and $\qquad x \ge 1$

There are no real numbers that satisfy the first conjunction, so the solution must be $\mathbf{1 \le x \le 5.}$

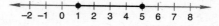

9.
(117)
$|x| + 2 > 5;\ D = \mathbb{Z}$

$\qquad |x| > 3$

$x > 3$ or $x < -3$

10.
(109)
$$x^{-1/3} - y^{2/5}$$
$$\underline{x^{-1/3} - y^{2/5}}$$
$$x^{-2/3} - x^{-1/3}y^{2/5}$$
$$\underline{\quad - x^{-1/3}y^{2/5} + y^{4/5}}$$
$$x^{-2/3} - 2x^{-1/3}y^{2/5} + y^{4/5}$$

11.
(108)
$$64a^3c^6 - 27x^9 = (4ac^2)^3 - (3x^3)^3$$
$$= (4ac^2 - 3x^3)(16a^2c^4 + 12ac^2x^3 + 9x^6)$$

12.
(104)
$$N = 3.0243\ 43\ 43\ \dots$$

$$100N = 302.43\ 43\ 43\ \dots$$
$$\underline{\quad N = \quad\ 3.02\ 43\ 43\ \dots}$$
$$99N = 299.41$$
$$N = \frac{29{,}941}{9900}$$

13.
(116)
$$\boxed{12\ |\ 11\ |\ 10\ |\ 9} = \mathbf{11{,}880}$$

14.
(98)
$$N = 3\frac{7}{10} + \frac{6}{7}\left(5\frac{2}{5} - 3\frac{7}{10}\right)$$
$$= \frac{37}{10} + \frac{6}{7}\left(\frac{54}{10} - \frac{37}{10}\right)$$
$$= \frac{37}{10} + \frac{6}{7}\left(\frac{17}{10}\right) = \frac{259}{70} + \frac{102}{70} = \mathbf{\frac{361}{70}}$$

15.
(90)
(a) $2x + 3y - 2z = -3$

(b) $x + 9y - 2z = 3$

(c) $3x + 6y + 2z = -1$

(a) $\quad 2x + 3y - 2z = -3$
-1(b) $\underline{-x - 9y + 2z = -3}$
(d) $\quad x - 6y \qquad\ = -6$

(a) $\quad 2x + 3y - 2z = -3$
(c) $\underline{3x + 6y + 2z = -1}$
(e) $5x + 9y \qquad = -4$

3(d) $\quad 3x - 18y = -18$
2(e) $\underline{10x + 18y = \quad -8}$
$\qquad 13x \qquad\ = -26$
$\qquad\qquad\ x = -2$

(d) $-2 - 6y = -6$
$\qquad -6y = -4$
$\qquad\quad y = \frac{2}{3}$

(b) $-2 + 9\left(\frac{2}{3}\right) - 2z = 3$
$\qquad\qquad\qquad -2z = -1$
$\qquad\qquad\qquad\ z = \frac{1}{2}$

$$\left(-2, \frac{2}{3}, \frac{1}{2}\right)$$

16.
(100)
$$y = x^2 - 2x - 3$$
$$y = (x^2 - 2x + 1) - 1 - 3$$
$$y = (x - 1)^2 - 4$$

From this we see:

(a) The graph opens upward.

(b) Axis of symmetry is $x = 1$.

(c) y-coordinate of vertex is -4.

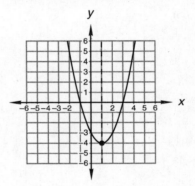

17.
(81)
$$\frac{-3i^4 + 5i^3}{6i^2 - 7i} = \frac{-3 - 5i}{-6 - 7i} \cdot \frac{-6 + 7i}{-6 + 7i}$$
$$= \frac{18 + 30i - 21i + 35}{36 + 49}$$
$$= \frac{53 + 9i}{85} = \mathbf{\frac{53}{85} + \frac{9}{85}i}$$

18.
(47)
$$\sqrt[3]{9\sqrt{3}} = [3^2(3)^{1/2}]^{1/3} = 3^{2/3}3^{1/6} = \mathbf{3^{5/6}}$$

19.
(118)
$$\log_5 (x + 3) + \log_5 4 = \log_5 36$$
$$\log_5 (x + 3)(4) = \log_5 36$$
$$4x + 12 = 36$$
$$4x = 24$$
$$x = \mathbf{6}$$

20.
(62)
$$6x - x^2 = 58$$
$$-58 = (x^2 - 6x + \quad)$$
$$9 - 58 = x^2 - 6x + 9$$
$$-49 = (x - 3)^2$$
$$\pm 7i = x - 3$$
$$\mathbf{3 \pm 7i = x}$$

TEST 31

1. T = tens digit
(107) U = units digit

(a) $10T + U = 7(T + U)$
   $10T + U = 7T + 7U$
   $3T = 6U$

(a') $T = 2U$

(b) $5U = T + 9$

Substitute (a') into (b) to get:
$5U = 2U + 9$
$3U = 9$
$U = 3$

(a') $T = 2(3) = 6$

Original number = **63**

2. $A_t = A_0 e^{kt}$
(115) $900 = 300 e^{k(6)}$
   $3 = e^{6k}$
   $\ln 3 = 6k$
   $\dfrac{\ln 3}{6} = k$

$A_{12} = 300 e^{[(\ln 3)/6](12)}$
$A_{12} = 300 e^{2 \ln 3}$
$A_{12} = 300 (e^{\ln 3})^2$
$A_{12} = 300 (3)^2$
$A_{12} = 300 (9)$
$A_{12} = $ **2700 bacteria**

3.
(120)

	Now	−10 Years	+10 Years
Linda:	L_N	$L_N - 10$	$L_N + 10$
Christy:	C_N	$C_N - 10$	$C_N + 10$

(a) $L_N - 10 = 3(C_N - 10)$

(b) $C_N + 10 = \dfrac{4}{7}(L_N + 10)$

(a') $L_N - 3C_N = -20$

(b') $-4L_N + 7C_N = -30$

$4(a')$ $4L_N - 12C_N = -80$
(b') $\underline{-4L_N + \;\; 7C_N = -30}$
   $-5C_N = -110$
   $C_N = $ **22 years old**

(c') $L_N - 3C_N = -20$
   $L_N - 3(22) = -20$
   $L_N = $ **46 years old**

4. Multiply both sides by $(x - 1)^2$:
(121) $x^2 - 1 \le 2(x^2 - 2x + 1)$
   $0 \le x^2 - 4x + 3$
   $0 \le (x - 3)(x - 1)$

$x = 1$ cannot be a solution since division by zero is not defined.

$(\text{Pos})(\text{Pos}) \ge 0$
$x - 3 \ge 0$ and $x - 1 > 0$
   $x \ge 3$ and $x > 1$

$(\text{Neg})(\text{Neg}) \ge 0$
$x - 3 \le 0$ and $x - 1 < 0$
   $x \le 3$ and $x < 1$

Thus, the solution is $x \ge 3$ and $x < 1$.

5. $|x - 2| \le 5; D = \mathbb{Z}$
(119) $x - 2 \le 5$ and $x - 2 \ge -5$
   $x \le 7$ and $x \ge -3$

6. $\log_6 (x - 4) + \log_6 7 = \log_6 21$
(118) $\log_6 [7(x - 4)] = \log_6 21$
   $7(x - 4) = 21$
   $x - 4 = 3$
   $x = $ **7**

7. $\log_{14} (x + 5) - \log_{14} (x - 5) = \log_{14} 11$
(118)
   $\log_{14} \dfrac{x + 5}{x - 5} = \log_{14} 11$
   $\dfrac{x + 5}{x - 5} = 11$
   $x + 5 = 11x - 55$
   $60 = 10x$
   $6 = x$

8. (a) $3x - y + 2z = 15$
(106)
(b) $3x + 2y + z = 4$

(c) $x - 2y + z = 10$

$2(a)$ $6x - 2y + 4z = 30$
(b) $\underline{3x + 2y + \;\; z = \;\; 4}$
(d) $9x \qquad\quad + 5z = 34$

(b) $3x + 2y + \;\; z = \;\; 4$
(c) $\underline{x - 2y + \;\; z = 10}$
(e) $4x \qquad\quad + 2z = 14$

(e') $z = 7 - 2x$

Substitute (e′) into (d) to get:

(d′) $9x + 5(7 - 2x) = 34$

$$-x = -1$$

$$x = 1$$

(e′) $z = 7 - 2(1) = 5$

(a) $3(1) - y + 2(5) = 15$

$$-y + 13 = 15$$

$$y = -2$$

(1, –2, 5)

9.
(73)
$$\frac{5 - \sqrt{6}}{\sqrt{6} + 3} \cdot \frac{\sqrt{6} - 3}{\sqrt{6} - 3} = \frac{5\sqrt{6} - 6 - 15 + 3\sqrt{6}}{6 - 9}$$

$$= \frac{8\sqrt{6} - 21}{-3} = \mathbf{7 - \frac{8}{3}\sqrt{6}}$$

10.
(113)
$$\frac{10^{7.80}}{10^{-6.42}} = 10^{7.80 + 6.42}$$

$$= 10^{14.22} \approx \mathbf{1.66 \times 10^{14}}$$

11.
(113)
$$(10^{5.44})^{2/5} = 10^{2.18} \approx \mathbf{149.66}$$

12.
(113)
$H^+ = 10^{-pH}$

$H^+ = 10^{-7.23}$

$H^+ \approx \mathbf{5.89 \times 10^{-8}}$ **moles per liter**

13.
(113)
$pH = -\log H^+$

$pH = -\log (3.28 \times 10^{-7})$

$pH \approx \mathbf{6.48}$

14.
(B)

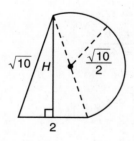

$$(\sqrt{10})^2 = H^2 + 1^2$$

$$10 = H^2 + 1$$

$$9 = H^2$$

$$3 = H$$

$$A = \frac{1}{2}(3)(2) + \frac{1}{2}(\pi)\left(\frac{\sqrt{10}}{2}\right)^2$$

$$= 3 + \frac{5}{4}\pi \approx \mathbf{6.93 \ m^2}$$

15.
(105)
$$6x^3 = 15x^2 + 36x$$

$$3x(2x^2 - 5x - 12) = 0$$

$$3x(2x^2 - 8x + 3x - 12) = 0$$

$$3x[2x(x - 4) + 3(x - 4)] = 0$$

$$3x(2x + 3)(x - 4) = 0$$

$3x = 0 \qquad 2x + 3 = 0 \qquad x - 4 = 0$

$x = 0 \qquad x = -\dfrac{3}{2} \qquad x = 4$

$\mathbf{0, -\dfrac{3}{2}, 4}$

16.
(104)
$$-12x + 5x^3 + 11x^2 = 0$$

$$x(5x^2 + 11x - 12) = 0$$

$$x(5x^2 + 15x - 4x - 12) = 0$$

$$x[5x(x + 3) - 4(x + 3)] = 0$$

$$x(5x - 4)(x + 3) = 0$$

$x = 0 \qquad 5x - 4 = 0 \qquad x + 3 = 0$

$\qquad\qquad x = \dfrac{4}{5} \qquad x = -3$

$\mathbf{0, \dfrac{4}{5}, -3}$

17.
(100)
$$y = -x^2 + 2x + 3$$

$$-y = (x^2 - 2x + 1) - 1 - 3$$

$$-y = (x - 1)^2 - 4$$

$$y = -(x - 1)^2 + 4$$

From this we see:

(a) The graph opens downward.

(b) Axis of symmetry is $x = 1$.

(c) y-coordinate of vertex is 4.

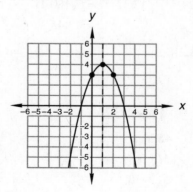

18.
(116)
$$\frac{1}{2} \times \frac{1}{12} \times \frac{1}{6} = \mathbf{\frac{1}{144}}$$

19. (a)
₍₁₂₂₎

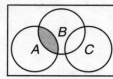

(b)

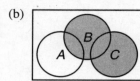

20.
₍₇₈₎

$A = 10 \cos 65° \approx 4.23$

$B = 10 \sin 65° \approx 9.06$

$A = 7 \cos 38° \approx 5.52$

$B = 7 \sin 38° \approx 4.31$

$\quad 4.23R + 9.06U$
$\underline{-5.52R + 4.31U}$
$\mathbf{-1.29R + 13.37U}$

$\tan \theta \approx \dfrac{13.37}{1.29}$

$\theta \approx 84.49°$

Subtract θ from 180° to get a second-quadrant angle:

$180° - 84.49° \approx 95.51°$

$F \approx \sqrt{(-1.29)^2 + (13.37)^2} \approx \sqrt{180.42}$
$\qquad\qquad\qquad\qquad \approx 13.43$

13.43 / 95.51°

TEST 32

1.
₍₁₂₀₎

	Now	+8 Years
Man:	M_N	$M_N + 8$
Son:	S_N	$S_N + 8$

(a) $M_N = 3S_N$

(b) $M_N + 8 = 2(S_N + 8) + 4$

Substitute (a) into (b) to get:

(b′) $3(S_N) + 8 = 2S_N + 20$

$\qquad\qquad S_N = $ **12 years old**

(a) $M_N = 3(12) = $ **36 years old**

2. $T = $ tens digit
₍₁₀₇₎

$U = $ units digit

$10T + U = $ original number

$10U + T = $ reversed number

(a) $T + U = 10$

(b) $\quad 10U + T = \dfrac{1}{2}(10T + U) - 13$

$\qquad 20U + 2T = 10T + U - 26$

(b′) $-8T + 19U = -26$

8(a) $\quad 8T + 8U = 80$
(b′) $\underline{-8T + 19U = -26}$
$\qquad\qquad\quad 27U = 54$
$\qquad\qquad\qquad U = 2$

(a) $T + 2 = 10$

$\qquad\quad T = 8$

Original number = **82**

3. Downstream: $(B + W)T_D = D_D$ (a)
₍₉₂₎

Upstream: $(B - W)T_U = D_U$ (b)

Since $T_D = 2T_U$, we substitute and get:

(a′) $12T_U + 2WT_U = 32$
2(b′) $\underline{12T_U - 2WT_U = 16}$
$\qquad 24T_U \qquad\quad = 48$
$\qquad\qquad T_U = 2$

(a′) $12(2) + 2W(2) = 32$

$\qquad\qquad 4W = 8$

$\qquad\qquad W = $ **2 mph**

4.
₍₁₂₃₎

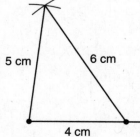

5.
(123)

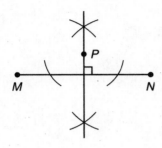

6.
(123)

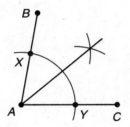

7.
(123)

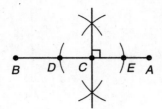

8. 1) $\overline{AD} \cong \overline{DC}$ and $\angle ADB \cong \angle CDB$ (given).
(124)

2) $\overline{DB} \cong \overline{DB}$ by reflexive property.

3) $\triangle ADB \cong \triangle CDB$ by SAS (1, 2).

4) $\overline{AB} \cong \overline{BC}$ by CPCTC.

9. 1) Circle P and $\overline{WX} \perp \overline{YZ}$ (given)
(125)

2) $\overline{PY}, \overline{PZ}$ are radii by definition of radii.

3) $\overline{PY} \cong \overline{PZ}$ because all radii of the same circle are congruent.

4) $\angle PXY$ and $\angle PXZ$ are right angles by definition of perpendicular.

5) $\triangle PXY, \triangle PXZ$ are right triangles by definition of right triangle.

6) $\overline{PX} \cong \overline{PX}$ by reflexive property.

7) $\triangle PXY \cong \triangle PXZ$ by HL (1, 2, 3, 4, 5, 6).

8) $\overline{YX} \cong \overline{ZX}$ and $\angle PXY \cong \angle PXZ$ by CPCTC.

9) $\overline{WX} \cong \overline{WX}$ by reflexive property.

10) $\triangle WXY \cong \triangle WXZ$ by SAS (8, 9).

11) $\angle Y \cong \angle Z$ by CPCTC.

10. 1) Circle O and $\angle BCD \cong \angle BAD$ (given)
(126)

2) $\overline{AD} \cong \overline{CD}$ because sides opposite equal angles in a triangle are always equal.

3) $\overline{OA}, \overline{OC},$ and $\overline{OD}$ are radii by definition of radii.

4) $\overline{OA} \cong \overline{OC} \cong \overline{OD}$ because all radii of the same circle are congruent.

5) $\overline{OD} \cong \overline{OD}$ by reflexive property.

6) $\triangle OCD \cong \triangle OAD$ by SSS (2, 4, 5).

7) $\angle ADB \cong \angle CDB$ by CPCTC.

8) $\angle ABD \cong \angle CBD$ because AA means AAA.

9) $\overline{OD} \perp \overline{CA}$ because two equal adjacent angles whose sum is 180° are right angles.

11. (a) S $\cup$ T = {2, 4, 5, 6, 7, 8, 10, 12}
(122)

(b) S $\cap$ T = {2, 8, 10}

12. $|x + 1| < 2$; $D = $ {Reals}
(119)

$x + 1 < 2$ and $x + 1 > -2$

$\quad x < 1$ and $\quad\quad x > -3$

13. Multiply both sides by $(x + 2)^2$
(121)

$x^2 - 4 > 3x^2 + 12x + 12$

$\quad 0 > 2x^2 + 12x + 16$

$\quad 0 > x^2 + 6x + 8$

$\quad 0 > (x + 4)(x + 2)$

(Pos)(Neg) < 0

$x + 4 > 0$ and $x + 2 < 0$

$\quad x > -4$ and $\quad\quad x < -2$

(Neg)(Pos) < 0

$x + 4 < 0$ and $x + 2 > 0$

$\quad x < -4$ and $\quad\quad x > -2$

There are no real solutions to the second conjunction. Thus, the solution is **$-4 < x < -2$.**

14. $A_6 = Pe^{rt}$
(115)

$A_6 = 3140e^{0.07(6)}$

$A_6 \approx \mathbf{\$4778.96}$

15. $\boxed{8 \mid 7 \mid 6 \mid 5}$ = **1680 signs**
(116)

16. $N = 1.023\ 23\ 23\ \ldots$
(104)

$$1000N = 1023.23\ 23\ 23\ \ldots$$
$$\underline{\quad 10N = \quad\ 10.23\ 23\ 23\ \ldots}$$
$$990N = 1013$$
$$N = \frac{\mathbf{1013}}{\mathbf{990}}$$

17. $3 \ln 0.0035 = 5x$
(115)

$$-16.96 \approx 5x$$
$$\mathbf{-3.39} \approx x$$

18. $y = x^2 + 6x + 8$
(100)

$$y = (x^2 + 6x + 9) - 9 + 8$$
$$y = (x + 3)^2 - 1$$

From this we see:

(a) The graph opens upward.

(b) Axis of symmetry is $x = -3$.

(c) y-coordinate of vertex is -1.

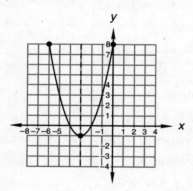

19. (a) $\log_b MN = \log_b M + \log_b N$
(122)

(b) $\log_b \dfrac{M}{N} = \log_b M - \log_b N$

(c) $\log_b M^N = N \log_b M$

20. $\log_3 (x + 5) - \log_3 (x - 3) = \log_3 9$
(118)

$$\log_3 \left(\frac{x + 5}{x - 3} \right) = \log_3 9$$
$$\frac{x + 5}{x - 3} = 9$$
$$x + 5 = 9x - 27$$
$$-8x = -32$$
$$x = \mathbf{4}$$

Test Forms

Instructions

Tests are an important component of the Saxon methodology. We believe that concepts and skills should be continually assessed. However, tests should be administered only after the concepts and skills have been thoroughly practiced. Therefore, we recommend that tests be administered according to the testing schedule printed on the reverse of this page.

Note: Optional student answer forms are located at the back of this booklet. These forms provide sufficient writing space for students to show all of their work.

Algebra 2

Testing Schedule

Test to be administered:	Covers material through:	Give after teaching:	
✓ Test 1	Lesson 2	Lesson 6	Q1
✓ Test 2	Lesson 6	Lesson 10	
Test 3	Lesson 10	Lesson 14	
Test 4	Lesson 14	Lesson 18	
Test 5	Lesson 18	Lesson 22	
Test 6	Lesson 22	Lesson 26	
Test 7	Lesson 26	Lesson 30	
Test 8	Lesson 30	Lesson 34	
Test 9	Lesson 34	Lesson 38	Q2
Test 10	Lesson 38	Lesson 42	
Test 11	Lesson 42	Lesson 46	
Test 12	Lesson 46	Lesson 50	
Test 13	Lesson 50	Lesson 54	
Test 14	Lesson 54	Lesson 58	
Test 15	Lesson 58	Lesson 62	
Test 16	Lesson 62	Lesson 66	
Test 17	Lesson 66	Lesson 70	Q3
Test 18	Lesson 70	Lesson 74	
Test 19	Lesson 74	Lesson 78	
Test 20	Lesson 78	Lesson 82	
Test 21	Lesson 82	Lesson 86	
Test 22	Lesson 86	Lesson 90	
Test 23	Lesson 90	Lesson 94	
Test 24	Lesson 94	Lesson 98	
Test 25	Lesson 98	Lesson 102	Q4
Test 26	Lesson 102	Lesson 106	
Test 27	Lesson 106	Lesson 110	
Test 28	Lesson 110	Lesson 114	
Test 29	Lesson 114	Lesson 118	
Test 30	Lesson 118	Lesson 122	
Test 31	Lesson 122	Lesson 126	
Test 32	Lesson 126	Lesson 129	

1. Find *m, n,* and *p.*

2. Find *x, y,* and *z.*

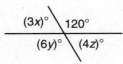

3. The complement of an angle is 37°. What is the measure of the angle?

4. Find the perimeter of this figure. All angles that look square are square. Dimensions are in inches.

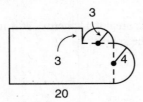

5. Find the area of the 60° sector of this circle. Dimensions are in inches.

6. The figure shown is the base of a cone whose altitude is 4 meters. What is the volume of the cone? Dimensions are in meters.

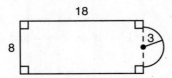

7. Find the area of the shaded region. Dimensions are in feet.

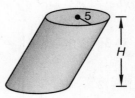

8. Find the area of the shaded region. Dimensions are in meters.

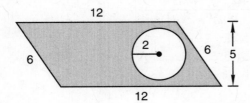

9. The volume of this circular cylinder is 750π in.^{3}. What is the height of the cylinder? Dimensions are in inches.

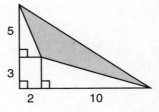

10. Find *a* and *b.*

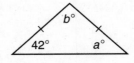

11. Find *y.*

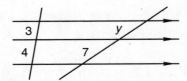

12. Find the volume and the surface area of a sphere whose radius is 8 centimeters.

Simplify. Write the answers with all variables in the numerator.

13. $\dfrac{(xy^2)^0 x^2 y}{x(y^{-3})^3}$

14. $\dfrac{(x^3 y^{-1})^{-2} z^{-2}}{(y^3 z y^{-2})^5}$

15. $\dfrac{x^3 y^2 z^{-2}}{(xw^0)^{-2} z^{-1} x^2 w^3}$

Simplify:

16. -3^{-5}

17. $\dfrac{1}{-3^{-3}}$

18. $-4^3 - [-5^0 - (3 - 5) - 4]$

19. $-|-3 - 5| - (-3)^2 - 3^2$

20. $-3[-6^0 - 2(6 - 8) - 2^3]$

1. Find y, A, and B.

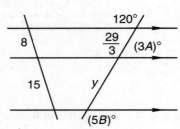

2. This figure is the base of a cone that is 12 cm tall. What is the volume of the cone? Dimensions are in centimeters.

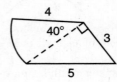

3. The measure of angle BCD is $70°$, as shown. Find x, y, P, R, and Q.

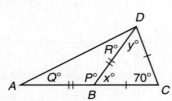

4. The surface area of a sphere is 38π in.^{2}. What is the radius of the sphere?

Evaluate:

5. $a - |b|a^2 - ba$ if $a = -3$ and $b = -4$

6. $m^2n(mn + n^2)$ if $m = -\dfrac{1}{2}$ and $n = \dfrac{1}{4}$

Simplify:

7. $2ab^{-1} + \dfrac{4a^2y^{-1}}{a} - \dfrac{7x^{-1}y}{y^{-2}}$

8. $-2mn + \dfrac{3mn^{-2}m^0}{n^{-3}} - \dfrac{5m^2m^{-1}m}{(m^{-1})^{-1}}$

9. $\dfrac{xb^0y(x^{-2}b^{-2})^2}{xb(by^0)xby}$

10. $\dfrac{4(a^{-2})^{-1}b^2a^4b^{-2}}{a^2aa^0a^{-5}(a^3)^2}$

11. $-5^0[-3^3 - 3(-3 - 2)](-4^0)$

12. $-3^{-3} - \dfrac{4}{-4^{-2}} - 2^0$

13. $-4^2 + (-2)^4 - 3^3 - |-4 - 4|$

Solve:

14. $0.003x + 0.6 = 2.88$

15. $4\dfrac{2}{3}x - 2\dfrac{1}{5} = 3\dfrac{1}{6}$

16. $-2 - 2^3 - 2(x - 2) = 2[(x - 2)2 - 2]$

17. Expand: $\dfrac{ab}{m}\left(\dfrac{-2m^{-2}}{ba} + \dfrac{3m}{a^{-1}b}\right)$

18. If x is the measure of an angle, $90 - x$ is the measure of the complement of the angle. If the measure of the angle equals four times the measure of its complement, what is the measure of the angle?

19. Find three consecutive odd integers such that three times the sum of the first and third is 18 greater than 4 times the second.

20. Research shows that 0.382 of the employees are totally loyal. If the corporation has 8000 employees, how many of them are not totally loyal?

Test 3   **SHOW YOUR WORK**   Name: _____

1. Thirty percent of what number is 480? Draw a diagram of the problem.

2. Sixty percent of the profit was used to pay the investors. If $15,000 was not used to pay the investors, what was the total amount of profit?

3. Find three consecutive even integers such that 3 times the first is 26 less than twice the sum of the last two.

4. The number that march in formation is $2\frac{3}{4}$ times the number that walk randomly. If 3300 march in formation, how many walk randomly?

5. In this figure $WX = WY$. Find a, b, and c.

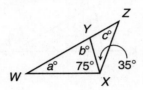

6. Find x.

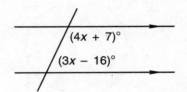

7. Find a, R, and T.

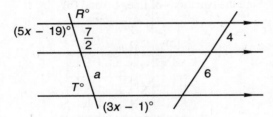

8. Use the Pythagorean theorem to find m.

Solve:

9. $2\frac{1}{4}n - \frac{5}{12} = \frac{3}{4}$

10. $-3(2x - 1) - 3^2 + 7^0 = -3(3x + 2x^0)$

Expand:

11. $\dfrac{p^{-2}c^3}{c^{-2}}\left(2p^2c - \dfrac{3pcp^2}{pc^2p}\right)$

12. $\dfrac{ac^2}{x}\left(\dfrac{3a^{-1}x}{c^2} - \dfrac{7cx}{a}\right)$

13. The area of a circle is 25π m^2. What is the circumference of the circle?

14. Graph on a rectangular coordinate system: $2x - y = 2$

15. Evaluate: $xy - x(y - x)$ if $x = -\dfrac{1}{2}$ and $y = -\dfrac{1}{3}$

16. Evaluate: $a^2c(a + c) - c^0$ if $a = -\dfrac{1}{3}$ and $c = -\dfrac{1}{2}$

Simplify:

17. $\dfrac{2^{-3}aaa^0(x^{-3}c)^{-2}}{(a^3c^{-1})^2acc^{-2}}$

18. $-4^0(-3^0 - 5^0 - |-3|) - (-3)(-6)$

19. $-\dfrac{1}{3^{-2}} - 2^3 - 2$

20. $\dfrac{5c^2y}{ac} + \dfrac{2c}{a} - \dfrac{4y^3c}{cya} - \dfrac{2c^3a}{a^2c^2}$

1. After the storm, the number of destroyed flowers was 240 percent more than before the storm. If 68,000 flowers were destroyed after the storm, how many had been destroyed before the storm?

2. A number is doubled and then the product is increased by 6. This sum is then multiplied by –3, and the result is 54 greater than three times the opposite of the number. What is the number?

3. Find three consecutive even integers such that 6 times the first is 16 less than four times the sum of the second and third.

4. Find c and d.

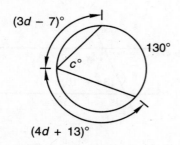

5. $\overline{XY}$ is a segment drawn along one side of a triangle. If $b = 150$ and $c = 70$, what is a?

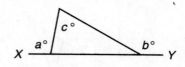

6. Find the area of this triangle. Dimensions are in centimeters.

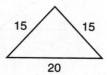

7. Find the equations of lines (a) and (b).

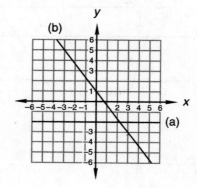

8. Add: $3 + \dfrac{r}{3t^2}$

9. Add: $\dfrac{a}{py} + y + \dfrac{a^2}{p^2}$

10. Find the distance between $(2, -3)$ and $(5, 6)$.

11. Graph (a) $x = 3$ and (b) $3x - 4y = 12$ on a rectangular coordinate system.

12. Solve: $0.3x - 0.3 - 0.03 = 0.33$

13. Solve: $-2^0(m^0 - 2) - 3(m - 3) = -2(m + 3^0)$

14. Expand: $\dfrac{3^{-2}x^{-2}}{y}\left(3x^2 y - \dfrac{9x^2}{y^2}\right)$

15. Simplify: $\dfrac{4^{-2}\, mmm^0 (m^{-3}p)^{-3}}{(m^3 p^{-1})^3 mpp^{-4}}$

16. Simplify by adding like terms: $-\dfrac{3pq}{r} + \dfrac{5p^4 p^{-3} r^{-1}}{q} - \dfrac{2r^{-1}p^{-2}}{p^{-3}q^{-1}}$

17. Evaluate: $a^2 - ya - y(a - y)$ if $y = -3$ and $a = \dfrac{1}{3}$

18. Simplify: $5^0 - \dfrac{75}{5^2} + (-4)^0 - 4^0 - 3\big[(-5 - 4^0) - (2 - 3)\big]$

19. Use substitution to solve: $\begin{cases} x + 3y = -3 \\ 3x - 5y = 19 \end{cases}$

20. Find the equation of the line that passes through $(-3, 5)$ and $(4, 2)$.

1. The formula requires 20 liters of alcohol to make 120 liters of the solution. How many liters of alcohol are needed to make 1560 liters of the solution?

2. Less than half of the students missed the chemistry demonstration. In fact only $\frac{3}{10}$ of the students missed the demonstration. If 21 students did not miss the demonstration, how many students did miss the demonstration?

3. Only 20 percent of the spectators left the contest discontented. If 6000 spectators left the contest contented, how many left discontented?

4. Use elimination to solve: $\begin{cases} 2x + y = 16 \\ 3x - 3y = 15 \end{cases}$

5. Add: $\dfrac{a}{c^2} - a - \dfrac{2c}{3a^2}$

6. Find d.

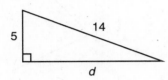

7. Graph: (a) $y = -2$ (b) $y = -3x$

8. Find the equation of the line that has a slope of $-\dfrac{2}{3}$ and passes through $(4, -3)$.

9. Solve for a, x, and y.

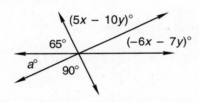

10. Find the area of the sector shown. Dimensions are in centimeters.

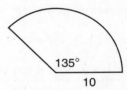

11. Expand: $\dfrac{mn^0}{-n^2 n^{-2}} \left(\dfrac{m}{n^3} - \dfrac{2m^3 n^2}{mn^3} \right)$

12. Solve for the unknown variables: $R_A T_A + R_B T_B = 300$, $R_A = 60$, $R_B = 12$, $T_B = T_A + 7$

13. Multiply: $(3x - 2)(4x^2 - 7x - 5)$

14. Divide $x^3 - 2x^2 - 1$ by $x - 3$ and check.

15. Solve: $-2[(-3)(-2 - x) + 2(x - 3)] = -2x$

16. Solve: $3\dfrac{1}{4}x - \dfrac{4}{5} = -\dfrac{3}{20}$

17. Simplify: $\dfrac{(-3a^0)^2 a^3 ac^2 c^0 c}{-3^{-3} a^2 cc^0 c^{-2} a}$

18. Simplify by adding like terms: $\dfrac{4m^3}{n^2} + 2mn^{-2}m^2 - 6\dfrac{mm}{n^2}$

19. Evaluate: $x - xy(y^2 - x^2)$ if $x = -\dfrac{1}{3}$ and $y = \dfrac{1}{2}$

20. Simplify: $-2^2[(-6 + 4) - |-2 + 5| - 3(-3^0 + (-3))]$

1. Mindy left at 8 a.m. and will arrive at 1 p.m.; Kelly left at 8 a.m. and will arrive at 2 p.m. How many miles is the trip if Kelly is traveling 10 miles per hour slower than Mindy?

2. Erasers cost 45 cents apiece. Notebooks cost 90 cents apiece. Karla spent $18 for a total of 28 erasers and notebooks. How many notebooks did she buy?

3. The ratio of two numbers is 8 to 5. The difference of the two numbers is 24. What are the numbers?

4. Use substitution to solve: $\begin{cases} 3x + y = -1 \\ 4x + 3y = 12 \end{cases}$

5. Divide $4x^3 + 3$ by $x - 1$ and check.

6. Add: $\dfrac{5x^3}{z} + 2b^2 - \dfrac{3x^2}{b^3 z}$

7. Find the equations of lines (a) and (b).

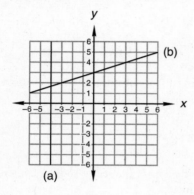

8. Find a and b.

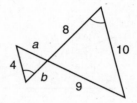

9. Find the equation of the line that passes through $(-3, -3)$ and is parallel to $x - 2y = -7$.

10. Simplify: $5\sqrt{3}(2\sqrt{8} - 3\sqrt{3})$

11. Simplify: $-6\sqrt{2}(5\sqrt{2} - 5\sqrt{10})$

12. Simplify: $\dfrac{(0.008 \times 10^{-6})(300 \times 10^5)}{(0.006)(400,000,000)}$

13. The triangle and the circle are tangent at three points, as shown. Find m and n.

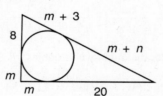

14. Solve: $0.4(x - 3) = 0.04(3x + 40)$

15. Solve: $-3[x - (-4 - 3^0) - 2] + [(x + 3)(-2)] = 2x$

16. Expand: $2x^{-2}y^3\left(\dfrac{x^2}{y^3} - 7x^{-2}y^{-3}\right)$

17. Simplify: $\dfrac{(a^2bc^{-1})^{-2}(ab^0c)^2}{(3a^2)^{-3}}$

18. Simplify: $-\dfrac{3xy^2}{x^{-1}} + 5xy^2x^{-1} - \dfrac{2x^3}{xy^{-2}}$

19. Simplify: $\dfrac{-2}{-2^{-3}} - \dfrac{3}{-3^{-2}}$

20. Evaluate: $ac^2(a^2 - c)$ if $a = -\dfrac{1}{3}$ and $c = \dfrac{1}{2}$

1. Angela made the trip by bus in 16 hours and the return trip by car in 20 hours. The bus traveled 10 miles per hour faster than the car. How far was the trip?

2. In his coin collection of dimes and quarters, Quang has 168 coins. If the collection has a value of $30, how many coins are dimes and how many are quarters?

3. Twice the number of apples is 11 less than three times the number of pears. If there are 12 pieces of fruit in all, how many are pears?

4. Forty percent of the signs are circular. If 3600 signs are noncircular, how many signs are there in all?

5. Divide $4x^4 - 3$ by $x - 1$ and check.

6. Simplify: $-3\sqrt{24} + 4\sqrt{81} + 3\sqrt{54}$

7. Simplify: $3\sqrt{6}(\sqrt{15} - 2\sqrt{6})$

8. Simplify: $\dfrac{12x^2z - 3x}{3x}$

9. Simplify: $\dfrac{(0.0005 \times 10^{-3})(40 \times 10^4)}{(100,000)(0.2 \times 10^{-15})}$

10. Add: $\dfrac{a^3c^2}{x} + x^3 - \dfrac{cx^2}{m}$

11. Solve this system by graphing, and then get an exact solution by using either substitution or elimination.

$$\begin{cases} 8x + y = 4 \\ 4x - 12y = 12 \end{cases}$$

12. Find the equation of the line that passes through the point $(-3, 8)$ and is parallel to $2x - 6y = 7$.

13. Find x, y, and z.

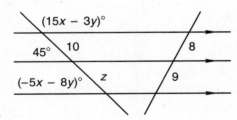

14. Find A, B, and C.

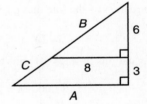

Solve:

15. $\dfrac{2x}{3} + \dfrac{3x - 4}{2} = 3$

16. $-4[x - (-2 - 3^0) - 2] + [(x - 2)(-3)] = 2x$

17. $\dfrac{3x}{4} - \dfrac{2x - 1}{5} = 3\dfrac{1}{2}$

Factor completely. Always factor the greatest common factor (GCF) as the first step.

18. $3ax^2 - 42a + 15ax$

19. $40cb - 2cbx^2 + 2bxc$

20. Evaluate: $a^0 - ax^0(x^2 - a) - \dfrac{1}{-2^{-2}}$ if $a = -\dfrac{1}{2}$ and $x = -3$

1. Karen left her house heading east, and her sister left two hours later heading west at a rate 5 mph faster than Karen. Six hours after Karen left, she and her sister were 600 miles apart. How fast was Karen moving?

2. The general expression for consecutive multiples of seven is $7N$, $7(N + 1)$, $7(N + 2)$, etc. Find three consecutive multiples of seven such that 3 times the first exceeds twice the third by 14.

3. The number of blooming pansies increased by 160 percent overnight. If they numbered 117 in the morning, how many pansies had bloomed before nightfall?

4. For every 220 students awarded scholarships, 5280 students had applied. If 869 students were awarded scholarships, how many had applied?

5. Divide $3x^3 - 5$ by $x - 1$ and check.

Factor completely. Always factor the GCF as the first step.

6. $20xz - 2x^2z - 42z$

7. $8xa + 42a - 2ax^2$

8. $63a^2xy + 9a^2x^2y + 108a^2y$

9. Simplify: $\dfrac{pm - 3p^2m^3}{pm}$

10. Simplify: $\dfrac{5}{2\sqrt{5}}$

11. Simplify: $3\sqrt{27} - 4\sqrt{108} - 2\sqrt{75}$

12. Simplify: $\dfrac{\frac{a + b}{x}}{\frac{c}{x}}$

13. Add: $\dfrac{3x}{x^2 + 3x - 10} - \dfrac{4}{x + 5}$

14. Solve this system by graphing, and then get an exact solution by using either substitution or elimination.

$$\begin{cases} 3x + 2y = 12 \\ 8x - 2y = 10 \end{cases}$$

15. Find the equations of lines (a) and (b).

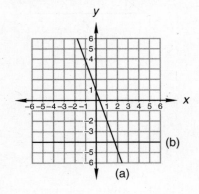

16. Find M, N, and P.

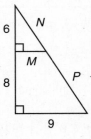

Solve:

17. $\dfrac{3x}{4} - \dfrac{2x - 3}{3} = 6$

18. $-3x - \dfrac{5^0 - x}{2} + \dfrac{x + 4^0}{3} = 4$

19. Evaluate: $a^2 - b^3(a - b)$ if $a = \dfrac{1}{3}$ and $b = -\dfrac{1}{2}$

20. The base of an 8-meter-tall right cylinder is shown. Find the lateral surface area of the cylinder. Dimensions are in meters.

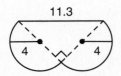

1. When Clint began his pursuit, Sue was already 60 miles ahead. If Sue travels at 40 mph and Clint travels at 60 mph, how many hours will it take Clint to catch up with Sue?

2. Find three consecutive even integers such that three times the sum of the first and third is 24 greater than 4 times the second.

3. Mel's coin jar contains 20 more nickels than dimes. The total value of the nickels and dimes is \$7. How many coins of each type are there?

4. Twenty percent of the ice melted. If 680 grams remain, how many grams melted?

5. Find the equation of the line that is perpendicular to $3x - 2y = 6$ and passes through $(1, 2)$.

Simplify:

6. $\dfrac{\dfrac{a}{c} - 1}{\dfrac{1}{c} + 3}$

7. $\dfrac{\dfrac{3}{p} - 2x}{p}$

8. $-4\sqrt{\dfrac{3}{10}} + 2\sqrt{\dfrac{10}{3}}$

9. $\dfrac{(50 \times 10^{-10})(0.0000024)}{8{,}000{,}000}$

10. $3\sqrt{63} - 5\sqrt{175} + \sqrt{252}$

11. $\dfrac{3a^2 b^3 + ab^2}{ab^2}$

12. Add: $\dfrac{2x}{x^2 + 4x - 12} - \dfrac{5}{x + 6}$

13. Multiply: $(a^2 + 2a + 5)(a^2 - a)$

Factor completely. Always factor the GCF as the first step.

14. $-2x^2 + 14x + 16$

15. $30x^2 a - 3ax^3 - 75ax$

Solve:

16. $\dfrac{2x - 5}{3} - \dfrac{x}{6} = 12$

17. $-3(x^0 - 2) + 4(-6^0 - x) = -2^0(2x - 5)$

18. Find an angle such that twice the supplement of the angle is 250° more than the complement of the angle.

19. Solve this sytem by graphing, and then get an exact solution by using either substitution or elimination.
$$\begin{cases} 6x + 4y = 2 \\ 10x - 8y = 32 \end{cases}$$

20. The figure shown is the base of a right cylinder that is 8 meters tall. Find the volume and the surface area of the cylinder. Dimensions are in meters. All angles that look square are square.

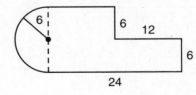

1. Linda headed east at 20 miles per hour on her bicycle at 9 a.m. Three hours later Shanna headed west from the same place, walking at 6 miles per hour. What time was it when they were 164 miles apart?

2. Audiocassette tapes cost $4 each, and compact discs cost $12 each. If the number of audiocassette tapes was 3 more than twice the number of compact discs and if a total of $92 was spent, how many audiocassette tapes and how many compact discs were purchased?

3. The chemical formula for sulfuric acid is H_2SO_4. If the weight of the acid is 1470 grams, how much does the sulfur weigh? (H, 1: S, 32; O, 16)

4. The number of students that enrolled in algebra was $2\frac{2}{3}$ times the number of textbooks. If 464 students enrolled in algebra, how many textbooks were there?

Simplify:

5. $(-8)^{-5/3}$

6. $\dfrac{1}{-64^{-2/3}}$

7. $3\sqrt{\dfrac{3}{13}} - 7\sqrt{\dfrac{13}{3}}$

8. $3\sqrt{50}\,(2\sqrt{18} - \sqrt{6}\,)$

9. Find x, y, and z.

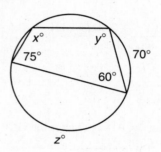

10. Find the equation of the line that passes through the points $(-4, -3)$ and $(6, -1)$.

11. Simplify: $\dfrac{\dfrac{3ac}{x^2} + \dfrac{2}{z^2}}{1 - \dfrac{4}{x^2z^2}}$

12. Simplify: $\dfrac{x^2 - 5x - 6}{x^2 - 2x - 3} \div \dfrac{x^2 - 11x + 30}{x^2 - x - 20}$

13. Simplify: $-2^2 - (-2)^2 - \left|2^2 - 3^2\right| - \dfrac{1}{-2^{-3}} - 3^0(-2^3 - 3^2)$

14. Add: $\dfrac{2}{x - 2} - \dfrac{2x - 1}{x^2 + x - 6} - \dfrac{2x}{x + 3}$

15. Multiply: $\dfrac{5c^{-2}m^{-2}}{a^2}\left(\dfrac{c^2m}{a^{-2}} - \dfrac{2c^2m^2a^2}{t}\right)$

16. Expand: $(x + 3)^3$

17. Find the distance between $(-5, -2)$ and $(4, -3)$.

Solve:

18. $\dfrac{3 - 2x}{3} + \dfrac{x}{5} = 2$

19. $3x(2 - 6^0) - 9x = 5x^0 - |-9 - 6| - 2(x - 4^0)$

20. $-4x = 12 - x^2$

1. Four times the larger number was 11 greater than seven times the smaller number. If the smaller number was 8 less than the larger number, what was the smaller number?

2. By taking a foreign language, students increase their vocabulary by 140 percent. If those who do not take a foreign language know 3640 words, how many words do the foreign language students know?

3. The chemical formula for potassium chlorate is $KClO_3$. If there were 6832 grams of this compound, how much did the chlorine weigh? (K, 39; Cl, 35; O, 16)

4. The total distance was 240 miles, and the first part was traveled at 20 mph. Then the speed was doubled for the second part. If the total trip took 7 hours, how many miles were traveled at 20 mph?

5. $WXYZ$ is a rhombus. Find A, B, and C.

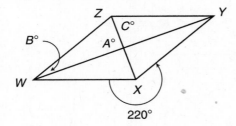

6. The area of this figure equals the sum of the areas of triangle AXP and the sector formed by arc ABC. Find the area of this figure. Dimensions are in inches.

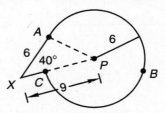

7. Find x and y.

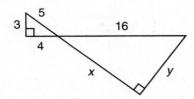

8. Estimate: $\dfrac{(8351)(30{,}804 \times 10^{-9})}{(0.0061 \times 10^{20})(0.00079)}$

9. Solve: $\dfrac{-4(x + 2)}{3} - \dfrac{x}{5} = 9$

10. Solve: $15x^2 = 34x - x^3$

11. Find the equation of the line that passes through $(-2, 5)$ and is perpendicular to $2x + 2y = 18$.

12. Find m: $\dfrac{by}{x} = c + \dfrac{a}{m}$

13. Use unit multipliers to convert 20 cubic miles to cubic inches.

Simplify:

14. $\dfrac{x^3 - 2x^2 - 15x}{x^2 - 3x - 10} \div \dfrac{x^3 + 21x + 10x^2}{x^2 + 8x + 7}$

15. $-27^{-4/3}$

16. $2\sqrt{6}(3\sqrt{8} - 5\sqrt{6})$

17. $4\sqrt{\dfrac{5}{7}} - 3\sqrt{\dfrac{7}{5}}$

18. $\dfrac{6a^3 - 6a^6}{6a^2}$

19. $\dfrac{\dfrac{a}{c} + 2}{5 - \dfrac{1}{c}}$

20. Evaluate: $-3x^{-1} - x^{-3} + x^{-2}$ if $x = -\dfrac{1}{4}$

1. The ratio of cassettes to compact discs is 10 to 3, and the number of cassettes is 10 less than 5 times the number of compact discs. How many cassettes are there?

2. Jamethan has 144 coins in his collection of nickels and quarters. If the coins have a total value of $20, how many are nickels and how many are quarters?

3. Maria finished with 180 percent of her original amount. If she finished with 7200, what was Maria's original amount?

4. Jake and Blake begin their journey from the same location at the same time. Jake's rate is 50 mph and Blake's rate is 60 mph. How many hours will it take for Blake to get 65 miles ahead of Jake?

Use trigonometric functions as necessary to find the missing parts of the triangles in problems 5 and 6.

5.
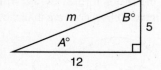

6.
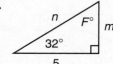

7. Find a: $\dfrac{2c}{a} - b = \dfrac{m}{x}$

8. Solve this system by graphing, and then get an exact solution by using either substitution or elimination.

$$\begin{cases} 3x - 4y = -12 \\ 6x + 2y = -3 \end{cases}$$

Solve:

9. $\left(x - \dfrac{2}{3}\right)^2 = 14$

10. $\dfrac{-2x + 3}{4} - \dfrac{3x + 5}{3} = 5$

11. $20x = x^3 - x^2$

Simplify:

12. $\dfrac{x^3 - 7x^2 - 18x}{x^2 - 81} \cdot \dfrac{x^2 + x - 30}{x^3 + 8x^2 + 12}$

13. $3\sqrt{\dfrac{3}{7}} - 2\sqrt{\dfrac{7}{3}} - \sqrt{189}$

14. $\sqrt[4]{mn^5}\,\sqrt[3]{m^5n^3}$

15. $\sqrt{8\sqrt{8}}$

16. $\dfrac{c - \dfrac{3x^2}{c}}{2c + \dfrac{z^2}{c}}$

17. $\dfrac{1}{-16^{-3/4}}$

18. Estimate: $\dfrac{(677{,}123)(5{,}134{,}692 \times 10^6)}{6023 \times 10^{-8}}$

19. Find the equation of the line that passes through $(3, -5)$ and is perpendicular to the line that passes through $(7, -5)$ and $(-2, -4)$.

20. Evaluate: $a^3 + ab^2 - ab$ if $a = -\dfrac{1}{3}$ and $b = -\dfrac{1}{2}$

1. Find three consecutive odd integers such that 6 times the sum of the first and second is 16 more than 8 times the third.

2. If a beaker contains 2720 grams of PH_3, what is the weight of the phosphorus? (P, 31; H, 1)

3. Mark and Amy live 560 miles apart. They are planning to meet somewhere between the two locations. If Mark travels at 20 mph and Amy travels at 50 mph and they leave at the same time, how far will Mark travel before they meet?

4. Use unit multipliers to convert 7 miles per hour to feet per second.

5. Find the equation of this line.

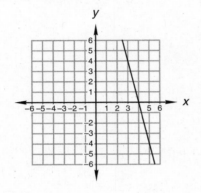

6. Find B and a.

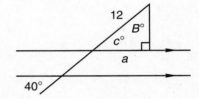

7. Find m: $\dfrac{2x}{m} + \dfrac{3a}{x} = \dfrac{5}{c}$

8. Solve by completing the square: $x^2 = 8 - 9x$

9. Solve by factoring: $189x = 3x^3 + 6x^2$

10. Solve: $9 = 3 + \sqrt{x - 2}$

11. Solve: $3x^0 - 4x(-1 - 3^0) = 5(x + 3)$

12. Simplify: $\sqrt[3]{27\sqrt{3}}$

13. Simplify: $\sqrt[3]{a^2c^4}\sqrt{a^5c^3}$

14. Simplify: $6\sqrt{\dfrac{2}{3}} - 8\sqrt{\dfrac{3}{2}} - \sqrt{294}$

15. Simplify: $\dfrac{x^3 - 6x + x^2}{-6 - 5x + x^2} \div \dfrac{2x^2 - 8x + x^3}{x^2 - 2x - 24}$

16. Find the distance between $(5, -2)$ and $(-3, -8)$.

17. Estimate: $\dfrac{(0.023 \times 10^{-8})(78{,}493 \times 10^5)}{18{,}000{,}000}$

18. Find c.

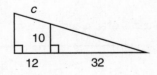

19. Find v, w, x, y, and z.

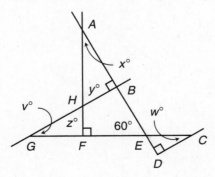

20. This figure is the base of a cone that is 18 meters high. Find the volume of the cone in cubic meters. Dimensions are in meters.

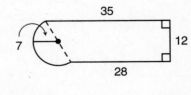

1. A pharmacist needs 25 liters of a solution that is 16% iodine. She has one solution that is 12% iodine and another solution that is 37% iodine. How many liters of each solution should the pharmacist use?

2. What percent by weight of $Na_2S_2O_3$ is sodium (Na)? (Na, 23; S, 32; O, 16)

3. The first part of the 660-mile journey was by car at 30 mph. The second part was by bus at 60 mph. If the total time was 14 hours, what was the distance traveled by bus?

Simplify:

4. $5i^5 - 3 + 2i - \sqrt{-36}$

5. $3i^3 + 5i - 3i^2 + \sqrt{-25} - 2$

6. $\sqrt[3]{25\sqrt{5}}$

7. $\sqrt[5]{m^2 p^7}\,\sqrt[3]{m^5 p}$

8. $2\sqrt{\dfrac{7}{13}} - 3\sqrt{\dfrac{13}{7}} - \sqrt{364}$

9. $\dfrac{1}{-27^{-2/3}}$

10. $\dfrac{\dfrac{r^2 c}{m^3} - c^2}{\dfrac{4}{m^3} - \dfrac{r}{m^3}}$

11. Solve by completing the square: $-7x = 5 - x^2$

Solve:

12. $\dfrac{5x - 3}{5} - \dfrac{4x + 2}{6} = 4$

13. $\sqrt{x^2 - 3x - 6} - 3 = x - 2$

14. Find the equation of the line that passes through $(2, -7)$ and is parallel to the line that passes through $(-4, 4)$ and $(-6, 6)$.

15. Find a.

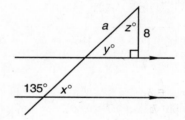

16. Use unit multipliers to convert 400 square kilometers to square miles.

17. Find x: $\dfrac{m}{c} - 3a + \dfrac{e}{x} = f$

18. Convert $8\underline{/64°}$ to rectangular form.

19. Find x in terms of a, b, and y.

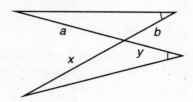

20. Find m, n, p, and q.

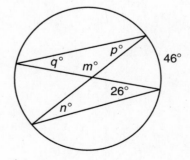

1. Sixteen liters of an ideal gas at a temperature of 500 kelvins has a pressure of 100 newtons per square meter. If the volume were increased to 20 liters and the temperature reduced to 400 kelvins, what would be the pressure?

2. Find three consecutive odd integers such that the product of the second and third is 55 less than 30 times the first.

3. A chemist has one solution that is 25% salt and another solution that is 10% salt. How many liters of each should the chemist use to make 120 liters of a solution that is 15% salt?

4. The carbon (C) in a container of C_6H_8NCl weighs 936 grams. How much does the chlorine (Cl) in the container weigh? (C, 12; H, 1; N, 14; Cl, 35)

5. Find a: $\dfrac{x - k}{p} - m = \dfrac{f}{a}$

6. Convert $8\underline{/310°}$ to rectangular form.

7. Use unit multipliers to convert 30 feet per second to meters per hour.

8. Complete the square: $3x^2 + 2x = 10$

9. Solve: $\sqrt{x^2 - 8x + 3} + 6 = x$

10. Solve: $\dfrac{5x - 2}{2} - \dfrac{x + 3}{3} = 2$

11. Find y in each figure.

(a)

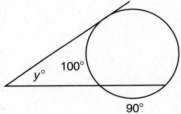

(b)

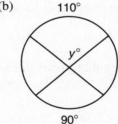

12. This figure is the base of a right cylinder that is 8 feet high. Find the volume of the cylinder. Dimensions are in feet.

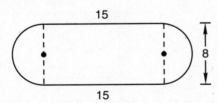

13. Find x.

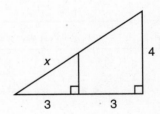

14. Solve this system by graphing, and then find the exact solution by using either substitution or elimination.

$$\begin{cases} 6x - 4y = 14 \\ 3x + 2y = -1 \end{cases}$$

Simplify:

15. $\sqrt[5]{27\sqrt{3}}$

16. $4i^2 - \sqrt{-16} + 3\sqrt{-9} - 4 + \sqrt{49}$

17. $\sqrt[4]{m^3 a^5}\, \sqrt[3]{m^5 a^4}$

18. $-16^{-5/4}$

19. $\dfrac{\dfrac{bc}{m} - \dfrac{em}{b^2}}{\dfrac{4}{b^2} - \dfrac{bc}{b^2 m}}$

20. $\dfrac{pqx^2 - 2pxq - 15qp}{-3x - 18 + x^2} \div \dfrac{-pqx + px^2q - 20qp}{-5x + x^2 - 6}$

1. The volume of an ideal gas is held constant. The initial pressure and temperature were 8000 pascals and 1500 kelvins. What would the final pressure be if the temperature were raised to 3270 kelvins?

2. How much pure alcohol should be added to 360 mL of a 30% alcohol solution to raise the concentration to 40%?

3. The number of correctly answered test questions varies directly as the number of attempted practice problems. Jason attempted 90 practice problems and answered 15 test questions correctly. How many test questions will Jason answer correctly if he attempts 102 practice problems?

4. The data points shown came from an experiment that involved tungsten (W) and iridium (Ir). Write the equation that expresses tungsten as a function of iridium: $W = m\mathrm{Ir} + b$

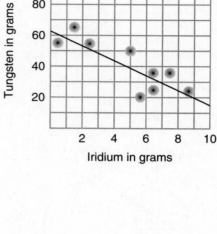

5. Solve: $\begin{cases} \dfrac{3}{8}x - \dfrac{1}{2}y = 1 \\ 0.2x + 0.6y = 4 \end{cases}$

6. Convert $14\underline{/320°}$ to rectangular form.

7. Convert $-3R - 4U$ to polar form.

8. Solve $3x^2 + 5x - 4 = 0$ by completing the square.

9. Solve $56x + x^3 = 15x^2$ by factoring.

10. Solve: $\dfrac{3x - 4}{2} - \dfrac{3}{5} = 7$

11. Find a: $\dfrac{4x}{2y + 3a} - p = \dfrac{c}{q}$

12. Use unit multipliers to convert 300 cubic inches to cubic meters.

13. Find a, x, and y.

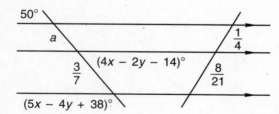

14. Find Z.

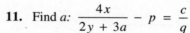

15. Find the equation of the line that passes through $(7, 5)$ and is perpendicular to $2x - 3y = 6$.

16. Solve this system by graphing, and then find an exact solution by using either substitution or elimination.

$$\begin{cases} 3x - y = -3 \\ x + y = 6 \end{cases}$$

Simplify:

17. $-3i^2 + 4i - 5\sqrt{-9} - 4 + 5i^3$

18. $3\sqrt{\dfrac{10}{11}} - 4\sqrt{\dfrac{11}{10}} - 5\sqrt{440}$

19. $\sqrt{20\sqrt{5}}$

20. $\dfrac{1}{(-32)^{-4/5}}$

1. The amount of aluminum varies inversely as the amount of tin. When there are 600 grams of aluminum, there are 40 grams of tin. How many grams of aluminum would there be if there were 120 grams of tin?

2. The initial pressure, volume, and temperature of a quantity of ideal gas were 450 newtons per square meter, 4 liters, and 300 kelvins, respectively. What would the pressure be if the temperature were increased to 500 kelvins and the volume were increased to 12 liters?

3. Two samples are available. One is 24% protein and the other is 12% protein. How much of each sample should be used to get 900 pounds that is 16% protein?

4. Write $-3R + 7U$ in polar form.

5. Add: $20\underline{/200°} + 50\underline{/40°}$

6. Solve: $\begin{cases} \dfrac{2}{3}x + \dfrac{2}{5}y = 18 \\ -0.1x - 0.2y = -5.5 \end{cases}$

7. Solve $4 = 2x - 3x^2$ by completing the square.

8. Solve: $-9^0 - 3^2 - 2^3(-3^0 - 3)x - 4x - 13x^0 = -x^0 + 3$

9. Find the surface area in square inches of the prism shown. Dimensions are in feet.

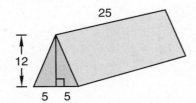

10. The data points shown came from an experiment involving hydrogen (H) and carbon (C). Write the equation that expresses hydrogen as a function of carbon: $H = mC + b$

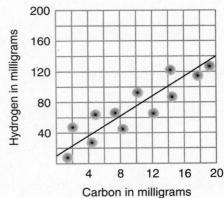

11. Find c: $\dfrac{a}{b + c} - d = \dfrac{e}{f}$

Simplify:

12. $\sqrt{-6}\sqrt{-7} - \sqrt{-25} - \sqrt{-3}\sqrt{-3} + 5i^7$

13. $(3i + 7)(2 - 5i)$

14. $\sqrt[4]{x^{1/3}\sqrt{x^5}}$

15. $\sqrt{8\sqrt{2}}$

16. $\dfrac{2x + 3}{x - 5} - \dfrac{5x - 4}{5 - x}$

17. Solve: $R_BT_B = 480$, $R_FT_F = 160$, $R_B = 6R_F$, $T_B + T_F = 6$

18. Use unit multipliers to convert 90 cubic inches per minute to cubic meters per second.

19. Simplify: $\dfrac{a}{m} + \dfrac{5}{3 + \dfrac{2a}{m}}$

20. Use similar triangles to find a and b.

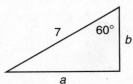

1. An ideal gas had an initial pressure of 0.03×10^9 atmospheres, an initial temperature of 7000 kelvins, and an initial volume of 0.007×10^{-2} cm^3. Find the final temperature if the final pressure is 3000×10^3 atmospheres and the final volume is 30,000 cm^3. Solve the equation for T_2 as the first step.

2. To arrive at 2000 liters of 60% saline solution, the attendant has to mix a 90% and a 40% saline solution. How many liters of each should be used?

3. Find three consecutive negative integers such that 3 times the product of the first and third exceeds twice the product of the first and second by 32.

Simplify:

4. $\dfrac{3}{3\sqrt{2} + 5}$

5. $6\sqrt{\dfrac{7}{5}} - 2\sqrt{\dfrac{5}{7}} - 3\sqrt{140}$

6. $\sqrt[4]{x^2 c^5} \sqrt[5]{xc^4}$

7. Add: $\dfrac{3}{x + 3} - \dfrac{2x + 1}{x^2 - 9}$

8. Add: $6\underline{/-60^\circ} + 6\underline{/215^\circ}$

9. Write $3R - 2U$ in polar form.

10. Solve: $R_M T_M = 240$; $R_Q T_Q = 240$; $R_M = 2R_Q$; $T_M = T_Q - 4$

Simplify:

11. $x + \dfrac{3b}{x + \dfrac{2}{b}}$

12. $(2i - 5)(4i - 3) - \sqrt{-5}\sqrt{-5} + 6i^2 - i^3$

13. Solve: $\begin{cases} \dfrac{1}{5}x - \dfrac{5}{2}y = -48 \\ 0.4x + 0.05y = 5 \end{cases}$

14. Solve by completing the square: $x + 6 = 2x^2$

15. Solve: $\dfrac{4x + 5}{3} + 2 = \dfrac{x}{7}$

16. Use unit multipliers to convert 300 cubic feet per minute to cubic yards per second.

17. Find x: $ay = b\left(\dfrac{c}{x + e} + \dfrac{3f}{h}\right)$

18. Find the equation of the line that passes through $(3, -5)$ and is parallel to $4x - 3y = 12$.

19. Use a calculator to simplify. Estimate first.

    (a) $\dfrac{373,402 \times 10^{10}}{97,376 \times 10^{-4}}$

    (b) $\sqrt[4.3]{207}$

20. Given: $m\widehat{AB} = 100^\circ$
    $\triangle ABC$ is isosceles
    $\overline{AC} = 10$
    $\overline{AB} = 8$
    Find $m\widehat{AC}$, $m\widehat{BC}$, and the area of $\triangle ABC$.

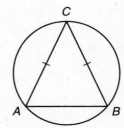

1. The volume of an ideal gas is held constant. The initial pressure and temperature were 1600 newtons per square meter and 240 kelvins. What would the final temperature be if the pressure were reduced to 1200 newtons per square meter?

2. Archimedes ran 3 times as fast as his challenger ran. In fact, he ran 60 miles in 3 hours less than it took his challenger to run 32 miles. How fast did each of them run? How long did they run?

3. Three times the number of blue marbles exceeds twice the number of red marbles by 18. Also, 5 times the number of blue marbles is 2 less than 6 times the number of red marbles. How many of each are there?

4. Begin with $ax^2 + bx + c = 0$ and complete the square to derive the quadratic formula.

5. Use the quadratic formula to solve: $2x^2 + 3 = -2x$

6. Find x in each figure.

   (a)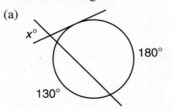

   (b)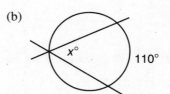

7. Find c: $xk = p\left(\dfrac{1}{cm_1} - \dfrac{1}{m_2}\right)$

8. Add: $\dfrac{2x + 1}{x + 4} + \dfrac{3x - 2}{(x + 4)(x - 1)}$

Simplify:

9. $\dfrac{3\sqrt{2} - 4}{\sqrt{3} - 2}$

10. $-2\sqrt{4\sqrt{2}}$

11. $\dfrac{-5^2}{-27^{-2/3}}$

12. $ax + \dfrac{a}{x - \dfrac{x}{a}}$

13. $(3i + 2)(2i - 3) - \sqrt{-16} - 2i^2 + i^3$

14. $\sqrt{a^3 x^0 y^2 a^{1/2} x^3 y}$

15. Find the equation of the line that passes through $(3, -1)$ and is perpendicular to the line that passes through $(-5, 2)$ and $(-1, 7)$.

16. Estimate the location of the line indicated by the data points. Write the equation that gives magnesium (Mg) as a function of calcium (Ca): $\text{Mg} = m\text{Ca} + b$.

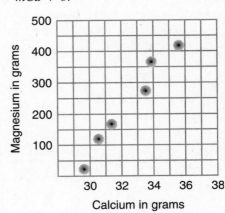

17. (a) Write $-3R + 3U$ in polar form.

    (b) Write $3\underline{/-60°}$ in rectangular form.

18. Solve $2x^2 + 3 = x$ by completing the square.

19. Solve: $3 + \sqrt{4x - 5} = 12$

20. Find Y.

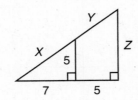

1. Ortiz drives twice as fast as his brother and therefore makes the 350-mile trip home in 5 hours less time. How fast does each drive, and how long does each travel?

2. A cornmeal feed is 20% protein, and a feed supplement is 60% protein. How much of each must be combined to get 3600 pounds of feed that is 30% protein?

3. Seven times the larger number is 8 more than four times the smaller. Also, three times the smaller is 19 less than twice the larger. What are the numbers?

Solve:

4. $\begin{cases} \dfrac{1}{5}x - \dfrac{1}{3}y = 1 \\ 0.03x - 0.7y = -3.75 \end{cases}$

5. $\begin{cases} x + y + z = 11 \\ 2x + y - z = 13 \\ x = 2y \end{cases}$

6. $\sqrt{x - 25} + \sqrt{x} = 5$

7. $\dfrac{4x - 3}{7} = 5 + \dfrac{x - 2}{3}$

8. The two forces act on a point as shown. Find the resultant force.

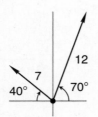

9. Use a calculator to simplify. Estimate first.

(a) $\dfrac{41{,}805 \times 10^3}{0.000395 \times 10^{-1}}$

(b) $\sqrt[2.3]{362}$

10. Find A.

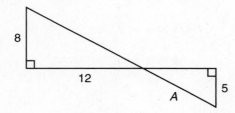

11. Find the area of this figure. Dimensions are in centimeters.

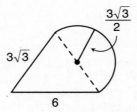

12. Solve $14 - 13x = -3x^2$ by using the quadratic formula.

13. Find b: $\dfrac{a}{m} = c\left(\dfrac{1}{x} + \dfrac{b}{y}\right)$

14. Add: $\dfrac{2x + 7}{x^2 - x - 6} - \dfrac{4x}{3 - x}$

Simplify:

15. $\dfrac{\sqrt{5} - 3}{\sqrt{5} - 2}$

16. $2\sqrt{\dfrac{5}{7}} + 3\sqrt{\dfrac{7}{5}} - 2\sqrt{140}$

17. $(5\sqrt{16} - 3)(2\sqrt{10} + 2)$

18. $ax^2 - \dfrac{a}{a + \dfrac{1}{ax}}$

19. $\sqrt{-16} - \sqrt{-3}\sqrt{-3} - \sqrt{-4}\sqrt{-4} + 3i - 3i^2 + 3i^3$

20. $\dfrac{-2^0(-4^2)}{-4^{3/2}}$

1. A's weight varies inversely as C's weight squared. When A weighs 20 kilograms, C weighs 10 kilograms. How much does A weigh when C weighs 5 kilograms?

2. The airplane could fly 2080 miles in 5 more hours than it took the sports car to drive 195 miles. The speed of the airplane was 4 times the speed of the sports car. What were the speed and time of each?

3. The ratio of two numbers is 8 to 5. Also, three times the larger number is 3 less than 5 times the smaller number. What are the numbers?

4. Use unit multipliers to convert 200 cubic centimeters per second to cubic feet per minute.

Solve:

5. $\sqrt{x - 95} + \sqrt{x} = 19$

6. $-4^2 - 2^0 + 3^2(2x - 1) = 5^0(x - x^0) + 9$

7. $\begin{cases} 3x + y + 4z = 15 \\ 2x - 2y + 3z = -3 \\ x - 2z = 0 \end{cases}$

8. $\begin{cases} \dfrac{1}{4}x - \dfrac{3}{2}y = -4 \\ 0.02x + 0.05y = 0.36 \end{cases}$

9. Find Y.

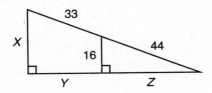

10. Find the equation of the line that passes through $(-2, -3)$ and whose slope is $\frac{2}{5}$.

11. Add: $-10\underline{/-35°} + 4\underline{/110°}$

12. Convert $6R - 11U$ to polar form.

13. Find the volume of a circular cone whose radius is 5 inches and whose height is 12 inches.

14. (a) Begin with $ax^2 + bx + c = 0$ and derive the quadratic formula by completing the square.
    (b) Solve $-9 + 4x = x^2$ by using the quadratic formula.

15. Add: $\dfrac{3x + 1}{x^2 - 16} + \dfrac{2x}{4 - x}$

Simplify:

16. $3\sqrt{7} + 5\sqrt{\dfrac{1}{7}} - 4\sqrt{28}$

17. $\dfrac{\sqrt{3} + 3}{\sqrt{3} - \sqrt{5}}$

18. $\dfrac{1}{2 + \dfrac{a}{2 + \dfrac{2}{x}}}$

19. $\dfrac{2 - 3i}{i - 5}$

20. Find y: $\dfrac{p}{m} = \dfrac{b}{m^2}\left(\dfrac{y}{a} + z\right)$

1. Bonnie drove 540 miles in three times the time it took Buffy to travel 135 miles. Buffy's speed was slower than Bonnie's by 15 miles per hour. What were the rate and time of each?

2. When the wildlife commission was finished, $\frac{1}{6}$ of the population of ducks had been observed. If the duck population numbers 26,340, how many ducks were observed?

3. A barrel contains 30 gallons of 40% antifreeze. How many gallons of an 80% antifreeze solution should be added to make the barrel contain 50% antifreeze?

Simplify:

4. $\dfrac{x^{4a}(y^b)^{3a}x^{a/4}}{y^{ba/3}}$

5. $\dfrac{(x^{a-3})^3}{x^{-3-2a}}$

6. $\dfrac{1}{x - \dfrac{a}{x - \dfrac{1}{a}}}$

7. Add: $\dfrac{2a-3}{a-c} - \dfrac{3a-1}{c-a}$

8. Solve: $\begin{cases} x + 2y + 4z = 3 \\ x - y - 2z = -3 \\ 3x + z = 0 \end{cases}$

9. Two forces are applied to an object as indicated. What is the resultant force on the object?

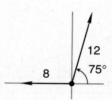

Simplify:

10. $\dfrac{4+5i}{i-2}$

11. $-\sqrt{-25} - \sqrt{2}\sqrt{-2} + \sqrt{-5}\sqrt{-5} + 3i - 2i^2 - 3i^3$

12. $5\sqrt{\dfrac{3}{5}} + 2\sqrt{60} - 2\sqrt{\dfrac{5}{3}}$

13. $\dfrac{-5 - 3\sqrt{5}}{2 - 4\sqrt{5}}$

14. Find a: $\dfrac{x+1}{m} - b = c\left(\dfrac{y}{a} + x\right)$

15. Solve: $\begin{cases} x^2 + y^2 = 25 \\ x - y = 1 \end{cases}$

16. Graph: $-x - 5 \not\geq -1$; $D = \{\text{Negative Integers}\}$

17. Use unit multipliers to convert 3 square miles per hour to square feet per minute.

18. Solve $2x^2 = 2x + 3$ by using the quadratic formula.

19. Solve $2x^2 + 2x = -3$ by completing the square.

20. Write equal ratios and find y in terms of a, b, and c.

1. After riding her bike at 20 miles per hour, Anisa walked her bike the rest of the way at 5 miles per hour. If the total trip was 70 miles and took a total of 5 hours, how far did she ride and how far did she walk?

2. Use the formula $PV = nRT$ to find the pressure of 23 liters of 0.642 mole of gas at a temperature of 800 K ($R = 0.0821$).

3. Haru is thinking of two numbers. The second number is 5 less than 3 times the first. Also, 10 times the first number is 5 less than 5 times the second. What two numbers is Haru thinking of?

4. Graph on a number line: $-1 \le x - 3 < 1$; $D = \{\text{Reals}\}$

5. Solve: $\begin{cases} BT_D + 5T_D = 25 \\ BT_D - 5T_D = 15 \end{cases}$

6. Solve: $\begin{cases} 2x - y + z = -2 \\ x + 2y + 2z = 3 \\ 2x - 2y + z = 0 \end{cases}$

7. Solve: $\sqrt{x} - 1 = \sqrt{x - 11}$

8. Find x in each figure.

   (a)
   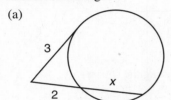

   (b)
   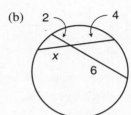

9. (a) Add: $3\underline{/30°} - 8\underline{/-240°}$

   (b) Write $3R - 7U$ in polar form.

10. Find the distance between $(-6, 2)$ and $(4, 6)$.

11. Use unit multipliers to convert 38 square kilometers to square miles.

12. Solve $5 - 2x = 6x^2$ by using the quadratic formula.

13. Find x: $\dfrac{a}{c} - y = m\left(\dfrac{1}{p} + \dfrac{k}{x}\right)$

14. Find the equation of the line that passes through $(-3, -2)$ and is parallel to $2y - 5x = 8$.

Simplify:

15. $\dfrac{(a^3)^{x+y} a^{x-2y} c^{2x}}{c^{3x/2}}$

16. $\dfrac{p}{x + \dfrac{xp}{1 + \dfrac{p}{x}}}$

17. $\dfrac{3 + 4i}{-2i - 5}$

18. $3i^2 + 4i - 2i^3 - \sqrt{-7}\sqrt{-7} + \sqrt{-3}\sqrt{-3}$

19. $\dfrac{5 + \sqrt{5}}{8 - 2\sqrt{5}}$

20. $\sqrt[3]{9\sqrt[5]{3}}$

1. The volume of an ideal gas is held constant. The initial pressure and temperature were 400 torr and 80 K. What would the final temperature be if the final pressure were 300 torr?

2. The number of chipmunks varies directly as the number of acorns in the forest. If there are 5 chipmunks when there are 100 acorns, how many are there when there are 30 chipmunks?

3. A paddleboat can travel 60 miles upstream in the same time it takes to travel 140 miles downstream. If the speed of the current is 10 miles per hour, what is the speed of the paddleboat in still water?

4. If $g(x) = x^2 + 2$ and $h(x) = x - 3$, what is $g(-1)$?

5. Use the discriminant to determine the kinds of numbers that satisfy the equation $5x = x^2 - 3$. Do not solve.

6. Find x in each figure.

   (a)

   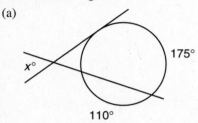

   $175°$

   $x°$

   $110°$

   (b)

   5

   4

   x

7. (a) Begin with $ax^2 + bx + c = 0$ and derive the quadratic formula by completing the square.

   (b) Solve $x = -2 - 3x^2$ by using the quadratic formula.

8. Solve: $\sqrt{x} = \sqrt{x + 12} - 2$

9. The two forces are applied to the point as indicated. What is the resultant force?

   7

   $60°$

   5

   $45°$

10. Graph: $\begin{cases} 2x + 6y \geq 18 \\ -3x + y < -2 \end{cases}$

11. Find x: $\dfrac{x + z}{a} = a\left(\dfrac{b}{m} + \dfrac{1}{c}\right)$

12. Solve: $\begin{cases} 2x - y + z = 7 \\ x + y - z = -4 \\ x + 3y - 2z = -11 \end{cases}$

13. Solve: $\begin{cases} x^2 + y^2 = 5 \\ x - 3y = 5 \end{cases}$

14. Graph on a number line: $x + 2 \not> 1$ or $x - 3 \not\leq -1$; $D = \{\text{Reals}\}$

Simplify:

15. $\dfrac{m^{x/4} n^3}{m^{2x/3} (n^a)^3}$

16. $\dfrac{2 - 3i}{4 - i}$

17. $5\sqrt{\dfrac{7}{20}} - 2\sqrt{\dfrac{20}{7}} - 3\sqrt{560}$

18. $\dfrac{5 + 2\sqrt{2}}{5\sqrt{2} - 2}$

19. $\dfrac{a}{b + \dfrac{1}{1 + \dfrac{c}{x}}}$

20. Use unit multipliers to convert 1200 cubic centimeters per hour to liters per second.

1. The number of boys varies directly as the number of girls and inversely as the number of teachers. When there are 24 boys and 16 girls, there are 2 teachers. How many girls are there when there are 72 boys and 6 teachers? Work the problem twice, once using the variation form and once using the equal ratio form.

2. The 888-mile trip to the mountains took three times as long as the 280-mile trip to the ocean because the speed to the ocean was 4 mph less than the speed to the mountains. What were the speed and time for each trip?

3. Determine the amounts of 20% and 50% salt solutions that should be mixed to obtain 300 gallons of 41% salt solution.

4. Solve: $\begin{cases} x - 4y = 12 \\ xy = 16 \end{cases}$

5. Which of the following sets of ordered pairs are functions?

   (a) (3, 2), (6, –1), (2, –1)

   (b) (3, 2), (2, –5), (3, –1)

   (c) (2, 3), (–3, 4), (–2, 3)

6. Graph: $\begin{cases} x + 2y > 6 \\ y \geq 3 \end{cases}$

7. Find the resultant of the two forces shown.

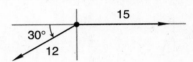

8. Graph on a number line:

   $-4 \leq x + 1 \not\geq 3;\ D = \{\text{Reals}\}$

9. Find c: $\dfrac{x}{c + m} = p\left(\dfrac{e}{y} + \dfrac{f}{z}\right)$

10. Use substitution to solve: $\begin{cases} 2x + 3y = 9 \\ 3x + 4y = 12 \end{cases}$

11. Find the number that is $\dfrac{2}{3}$ of the way from $\dfrac{3}{4}$ to $3\dfrac{5}{12}$.

12. Solve: $\dfrac{x - 1}{5} + 4 = \dfrac{x - 2}{3}$

13. Solve $-3x^2 + 2x - 7 = 0$ by completing the square.

14. Use unit multipliers to convert 5000 meters per hour to feet per second.

Simplify:

15. $\dfrac{(a^{3c})^{1/2}\, a^{3c}}{x^{c/3}}$

16. $\dfrac{5i^2 - 3i^3}{i^3 + 3i^2}$

17. $-(-\sqrt{-16}) - \sqrt{-5}\sqrt{-5} + 3 - 3i^3 - 3i^2 + 3i$

18. $\dfrac{xy}{x + \dfrac{y}{x + \dfrac{1}{xy}}}$

19. $\sqrt[5]{9\sqrt[4]{3}}$

20. $4\sqrt{\dfrac{5}{7}} + 3\sqrt{\dfrac{7}{5}} - 2\sqrt{315}$

1. The television retails for \$375. What is the purchase price of the television if it has been marked up 25% of the retail price?

2. A barge can travel 45 miles downstream in the same time it takes to travel 30 miles upstream. If the speed of the barge in still water is 15 miles per hour, what is the speed of the current?

3. Find three consecutive integers such that twice the product of the first and second exceeds the square of the third by 4.

4. Graph on a number line: $|x| + 3 \not> 7$; $D = \{\text{Integers}\}$

5. Find the number that is $\dfrac{3}{4}$ of the way from $2\dfrac{1}{3}$ to $4\dfrac{1}{6}$.

6. Use substitution to solve: $\begin{cases} 4x - 2y = -28 \\ 2x + 3y = -6 \end{cases}$

7. Solve: $\begin{cases} x^2 + y^2 = 36 \\ 2x^2 - y^2 = -9 \end{cases}$

8. Solve: $\begin{cases} x - 3y - z = 7 \\ 2x + y - 2z = 0 \\ x - y + z = 1 \end{cases}$

9. Find $gh(-4)$ if $g(x) = x + 4$; $D = \{\text{Reals}\}$ and $h(x) = x^2 - 6$; $D = \{\text{Integers}\}$.

10. Graph: $\begin{cases} x \geq 2 \\ 4x - 5y > 15 \end{cases}$

11. Complete the square as an aid in graphing: $y = -2x^2 + 8x - 6$

12. Graph on a number line: $x - 2 \not> 4$ or $x - 1 > 7$; $D = \{\text{Integers}\}$

13. Write $-3R - 7U$ in polar form.

14. Find c: $\dfrac{a}{c} = mn + \dfrac{a}{x + y}$

Simplify:

15. $\dfrac{cx}{1 - \dfrac{cx}{c - \dfrac{x}{c}}}$

16. $\dfrac{x^{c/4} y^{5c}}{x^{3c} y^{2c/3}}$

17. $\dfrac{4 - 3\sqrt{3}}{2\sqrt{3} - 4}$

18. $\sqrt[3]{x^5 y} \sqrt{x^2 y^4}$

19. $\dfrac{3i - 5}{-i^3 + 3i^2}$

20. $3i^9 - \sqrt{-4}\sqrt{-4} + \sqrt{5}\sqrt{-5} - \sqrt{-9}$

1. Henri purchased a stereo from a wholesaler for $400 and then sold it in his electronics warehouse for $450. What was his markup as a percentage of his purchase price?

2. The speedboat traveled at a speed that was twice as fast as the yacht. The speedboat traveled 240 miles in one hour less than it took the yacht to travel 150 miles. What were the speed and time for both the speedboat and the yacht?

3. Divide $a^3 + c^3$ by $a + c$.

4. Which sets of ordered pairs designate functions?

   (a) (1, 4), (2, 5), (3, 6), (1, 1)

   (b) (−1, −3), (0, 0), (1, 0), (−3, 2)

   (c) (0, 5), (5, 0), (4, 2), (3, 1)

   (d) (3, 2), (5, 2), (7, 2), (2, 2)

5. Find the resultant force of the vectors shown.

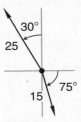

6. Find the number that is $\frac{1}{3}$ of the way from $1\frac{1}{4}$ to $3\frac{5}{12}$.

7. Solve: $\begin{cases} \dfrac{2}{3}x - \dfrac{2}{5}y = 2 \\ 0.15x + 0.006y = 0.93 \end{cases}$

8. Solve: $\begin{cases} 5x + y = 8 \\ xy = 3 \end{cases}$

9. Complete the square as an aid in graphing:
   $y = -x^2 - 4x - 6$

10. Graph: $\begin{cases} 3x + 2y < 8 \\ y < -3 \end{cases}$

11. Graph on a number line: $-|x| + 3 \not\geq 7$; $D = \{\text{Reals}\}$

12. Solve: $\begin{cases} 3x + 2y = 18 \\ 2x - z = 4 \\ y + 2z = 21 \end{cases}$

13. Solve: $\sqrt{x - 12} = 6 - \sqrt{x}$

Simplify:

14. $2\sqrt{\dfrac{5}{7}} - 3\sqrt{\dfrac{7}{5}} + \sqrt{140}$

15. $\dfrac{a^{2m}(a^{m/2 + 1})^4 c^n}{c^{n/2} a^{m/3}}$

16. $\dfrac{3 - 2i - 4i^3}{i + 2i^2 - 5i^3}$

17. Use similar triangles to find:

   (a) w and x

   (b) y and z

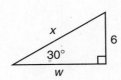

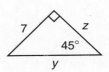

18. Show that each number is rational by writing it as a quotient of integers.

   (a) $0.006\overline{01}$

   (b) $0.0\overline{32}$

Solve by factoring:

19. $3x^2 + 5x - 2 = 0$

20. $3x^2 - 13x + 12 = 0$

1. The boat can travel 112 miles downstream in 4 hours but requires 6 hours to travel 48 miles upstream. What is the speed of the boat in still water, and what is the speed of the current?

2. The cylinder contains 1224 grams of the compound $CaSO_4$. What is the weight of the oxygen (O) in the compound? (Ca, 40; S, 32; O, 16)

3. The sum of the digits of a two-digit counting number is 15. If the digits are reversed, the new number is 9 more than the original number. What is the original number?

4. Farmer Brown purchased a tractor for $25,000 and sold it for $30,000. What was the markup as a percentage of the purchase price, and what was the markup as a percentage of the selling price?

5. Solve: $\begin{cases} x + 2y + z = 7 \\ 3x - y + z = -12 \\ 4x + 3y - 2z = 9 \end{cases}$

6. Solve: $\begin{cases} \dfrac{2}{3}x - \dfrac{2}{5}y = 2 \\ 0.03x + 0.04y = 0.67 \end{cases}$

7. Show that $3.6\overline{123}$ is a rational number by writing it as a fraction of integers.

8. Complete the square as an aid in graphing:
    $y = -x^2 + 4x + 1$

9. Graph: $\begin{cases} 4x < -12 \\ 2x + y \geq -4 \end{cases}$

10. Solve by factoring: $14x^3 = 42x - 7x^2$

11. Solve: $\sqrt{x} + 2 = \sqrt{x + 12}$

12. Factor: $27x^3y^6 - a^9c^{12}$

13. Expand: $(a^{3/2} + c^{1/4})^2$

14. Find $fg(2)$ if $f(x) = (x - 1)^2$; $D = \{$Reals$\}$ and $g(x) = x + 3$; $D = \{$Integers$\}$.

15. Graph on a number line: $x^2 - 5x \geq -6$; $D = \{$Reals$\}$

16. Use unit multipliers to convert 400 liters per minute to milliliters per second.

17. Solve: $\begin{cases} x + 3y = 7 \\ xy = 2 \end{cases}$

Simplify:

18. $\dfrac{4i - 3i^2 - 2}{\sqrt{-25} - \sqrt{-3}\sqrt{-3}}$

19. $\sqrt[5]{16\sqrt{2}}$

20. Write $-8R + 17U$ in polar form.

1. Albert raced 4 times as fast as Bernhard. In fact, Albert raced 48 miles in 3 hours less than it took Bernhard to race 30 miles. How fast did each of them race? What were the times of each racer?

2. There are 27 nickels, dimes, and quarters in the drawer with a value of $3. How many coins of each type are there if there are three times as many nickels as there are dimes?

3. The sum of the digits of a two-digit counting number is 10. If the digits are reversed, the new number is two less than three times the original number. What is the original number?

Graph each solution on a number line:

4. $x^2 + x \le 6$; $D = \{\text{Reals}\}$ 5. $x^2 + 7x < -10$; $D = \{\text{Reals}\}$

6. $|x| + 3 \not> 8$; $D = \{\text{Reals}\}$

7. Multiply: $(x^{1/4} + y^{1/2})(x^{-1/2} - y^{-1/4})$

8. Factor: $8a^3c^9 - 64x^{12}y^6$

9. Find x:

   (a) $x = \log 3.2041$ (b) $\ln x = 6.417$ (c) $e^x = 14.3$ (d) $\log x = 2.735$

10. Show that $0.003\overline{20}$ is a rational number by writing it as a quotient of integers.

11. Complete the square as an aid in graphing:
    $y = x^2 + 4x - 2$

12. Which region of the graph satisfies this system of inequalities?

    $$\begin{cases} x^2 + y^2 \le 9 & \text{(circle)} \\ y > 2x - 1 & \text{(line)} \end{cases}$$

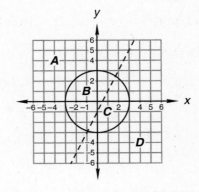

13. Solve: $\begin{cases} 1\frac{1}{5}x + \frac{1}{2}y = 22 \\ 0.1x + 0.5y = 11 \end{cases}$ 14. Solve: $\begin{cases} 4x + 2y = 10 \\ -2x + z = -1 \\ x + 3z = 11 \end{cases}$ 15. Solve: $\begin{cases} x^2 + y^2 = 34 \\ 2x - y = 1 \end{cases}$

16. Find the number that is $\frac{3}{4}$ of the way from $2\frac{5}{6}$ to $8\frac{5}{12}$.

Simplify:

17. $\dfrac{4i^2 - 3i}{-\sqrt{-5}\sqrt{-5} + 2i^3}$ 18. $\dfrac{\sqrt{2} - 5}{5\sqrt{2} + 3}$

19. Solve $3x = 2x^2 - 2$ by factoring.

20. Solve $-4x = 2x^2 - 7$ by using the quadratic formula.

1. The current in the river flows at 3 miles per hour. The boat can travel 24 miles downstream in one-half the time it takes to travel 12 miles upstream. What is the speed of the boat in still water?

2. The value of a group of 48 nickels, dimes, and quarters is $8.25. If there are twice as many quarters as dimes, how many coins of each kind are there?

3. Find three consecutive multiples of 5 such that the sum of the squares of the first two is 125 greater than the square of the third.

4. Use the [LN] key to express each number as a power. Then use the rules of logarithms to find the answer.

$$\frac{(53{,}671 \times 10^6)(23{,}241 \times 10^{-5})}{62{,}431}$$

5. Find the pH if $H^+ = 2.68 \times 10^{-3}$ mole per liter. ($pH = -\log H^+$ or $10^{-pH} = H^+$)

6. Julie deposited $13,000 into an account that paid $8\frac{1}{2}$ percent interest compounded continuously. How much money did she have after 9 years? ($A_t = Pe^{rt}$)

7. The figure shown is the base of a right cylinder 3 meters high. Find the volume of the cylinder in cubic meters. Dimensions are in meters.

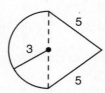

Graph the solutions:

8. $\{x \in \mathbb{R} \mid x^2 + 5 \le 6x\}$

9. $\{x \in \mathbb{Z} \mid |x| + 2 > 5\}$

10. Expand: $(x^{-1/3} - y^{2/5})^2$

11. Factor: $64a^3c^6 - 27x^9$

12. Show that $3.02\overline{43}$ is a rational number by writing it as a fraction of integers.

13. There are 4 infield positions on a baseball team. There are 12 players. How many ways can these players play the 4 positions?

14. Find the number that is $\frac{6}{7}$ of the way from $3\frac{7}{10}$ to $5\frac{2}{5}$.

15. Solve: $\begin{cases} 2x + 3y - 2z = -3 \\ x + 9y - 2z = 3 \\ 3x + 6y + 2z = -1 \end{cases}$

16. Complete the square as an aid in graphing:
$y = x^2 - 2x - 3$

Simplify:

17. $\dfrac{-3i^4 + 5i^3}{6i^2 - 7i}$

18. $\sqrt[3]{9\sqrt{3}}$

19. Solve for x: $\log_5 (x + 3) + \log_5 4 = \log_5 36$

20. Solve by completing the square: $6x - x^2 = 58$

1. A two-digit counting number has a value that is 7 times the sum of its digits. If 5 times the units digit is 9 more than the tens digit, what is the number?

2. The number of bacteria increased exponentially. At first there were 300. Six hours later there were 900. How many bacteria will there be at the end of 12 hours?

3. Ten years ago Linda was three times Christy's age then. Ten years from now Christy's age will be $\frac{4}{7}$ of Linda's age then. How old are they now?

Graph the solution on a number line:

4. $\dfrac{x + 1}{x - 1} \le 2;\ D = \{\text{Reals}\}$

5. $\left\{ x \in \mathbb{Z} \mid |x - 2| \le 5 \right\}$

Solve for x:

6. $\log_6 (x - 4) + \log_6 7 = \log_6 21$

7. $\log_{14} (x + 5) - \log_{14} (x - 5) = \log_{14} 11$

8. Solve: $\begin{cases} 3x - y + 2z = 15 \\ 3x + 2y + z = 4 \\ x - 2y + z = 10 \end{cases}$

9. Simplify: $\dfrac{5 - \sqrt{6}}{\sqrt{6} + 3}$

Use logarithms as required to perform the following operations. Begin by writing each number as an exponential expression whose base is 10.

10. $\dfrac{6.271 \times 10^7}{0.0038 \times 10^{-4}}$

11. $(274{,}000)^{2/5}$

12. Find the concentration of hydrogen ions (H^+) in moles per liter when the pH of the liquid is 7.23. ($pH = -\log H^+$ or $10^{-pH} = H^+$)

13. Find the pH of the solution when the concentration of hydrogen ions (H^+) in moles per liter is 3.28×10^{-7}. ($pH = -\log H^+$ or $10^{-pH} = H^+$)

14. Find the area of this figure. Dimensions are in meters.

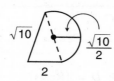

15. Solve $6x^3 = 15x^2 + 36x$ by factoring.

16. Solve $-12x + 5x^3 + 11x^2 = 0$ by factoring.

17. Complete the square as an aid in graphing: $y = -x^2 + 2x + 3$

18. Anna rolls a pair of dice three times. What is the probability she will get an odd sum on the first roll, a total of 4 on the second roll, and a total greater than 9 on the third roll?

19. Designate the area represented by
   (a) $A \cap B$ (b) $B \cup C$

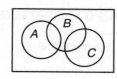

20. Find the resultant vector of the two vectors shown.

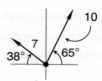

Outcome of second die

	1	2	3	4	5	6
1	2	3	4	5	6	7
2	3	4	5	6	7	8
3	4	5	6	7	8	9
4	5	6	7	8	9	10
5	6	7	8	9	10	11
6	7	8	9	10	11	12

Outcome of first die

Test 32 **SHOW YOUR WORK** **Name:** ______________________

1. A man is 3 times as old as his son. In 8 years he will be 4 years older than twice his son's age then. How old are they now?

11.

12.

13.

14.

15.

16.

17.

18.

19.

20.